AIChE/A&WMA Joint Workshop: Shale Oil and Gas E&P 2013

Water Challenges and Opportunities

Pittsburgh, Pennsylvania, USA
31 October – 2 November 2012

ISBN: 978-1-5108-1597-1

Printed from e-media with permission by:

Curran Associates, Inc.
57 Morehouse Lane
Red Hook, NY 12571

Some format issues inherent in the e-media version may also appear in this print version.

Printed by Curran Associates, Inc. (2015)

For permission requests, please contact AIChE
at the address below.

AIChE
120 Wall Street, FL 23
New York, NY 10005-4020

Phone: (800) 242-4363
Fax: (203) 775-5177

www.aiche.org

Additional copies of this publication are available from:

Curran Associates, Inc.
57 Morehouse Lane
Red Hook, NY 12571 USA
Phone: 845-758-0400
Fax: 845-758-2634
Email: curran@proceedings.com
Web: www.proceedings.com

TABLE OF CONTENTS

Hydraulic fracturing
demands

TECHNOLOGY
& INNOVATION

Apache

CAL COOPER

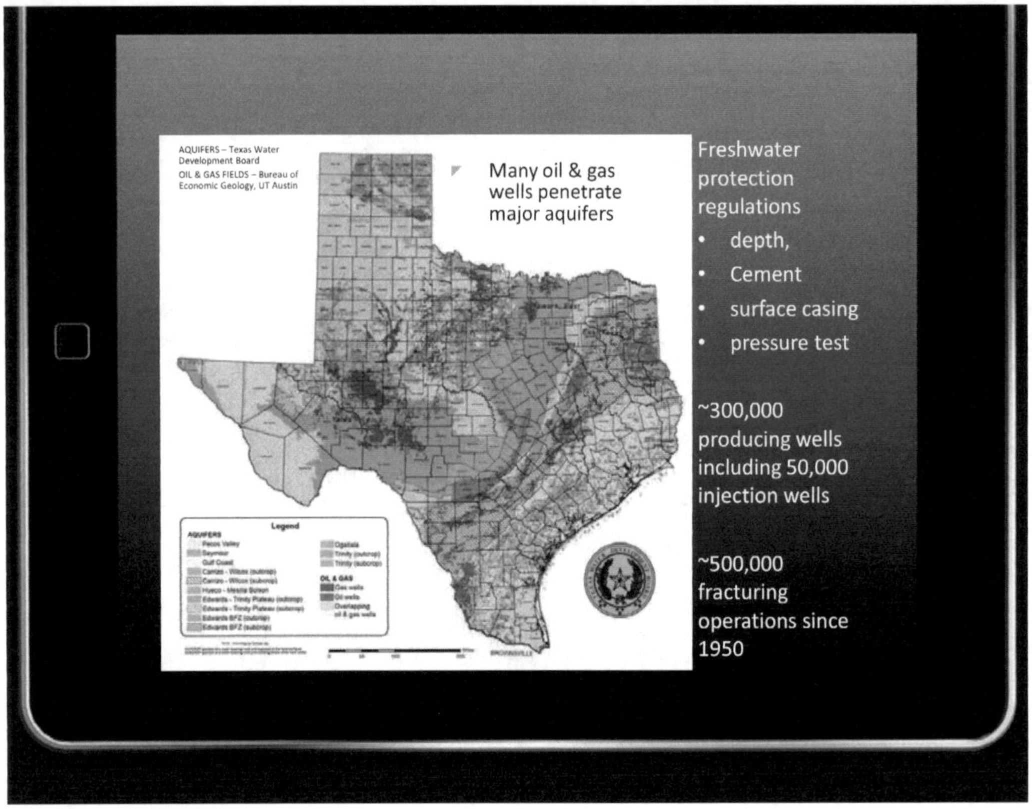
AQUIFERS – Texas Water Development Board
OIL & GAS FIELDS – Bureau of Economic Geology, UT Austin
Many oil & gas wells penetrate major aquifers
Legend
Freshwater protection regulations
• depth,
• Cement
• surface casing
• pressure test
~300,000 producing wells including 50,000 injection wells
~500,000 fracturing operations since 1950

iWater
iDrill
iChem
iPipe
iFrac
Process
IDEA
NEW
TECHNOLOGY
iInnovate
iEnable
4

INNOVATION DRIVES OIL & GAS BUSINESS

ACCESS TO ENERGY

MAJOR CAPITAL

STRONG COMPETITION

PERFORMANCE

Necessity....

5

OUR LABORATORY

6

WATER TREATMENT

SOURCE WATER

PRODUCED WATER RECYCLE

WATER DISPOSAL

Chemical and physical treatment

7

DRILLING FLUIDS

ISOLATE REACTIVE SURFACES

"INTELLIGENT" SEALS FOR FLUID LOSSES

LUBRICITY – FOR PIPE AND ROCK

Nano inspired innovations

8

CEMENT

- SOLIDIFY REMAINING DRILLING FLUIDS
- ELASTIC DUCTILE CEMENT FOR HIGH P
- IMPROVED BONDING IN SHALE

9

FRAC ADDITIVES

- BIOCIDES
- FRICTION REDUCERS
- GELLING AGENTS
- SURFACTANTS

Reaction regulators, enzymes,

10

PROPPANTS

SAND -NATURAL
- TREATED
- RESIN COATED

MANUFACTURED
- BEADS
- CARBO-CERAMICS
- SPECIALTY NANO

Pressure, Temperature, pH, chemical interaction with reservoir, flow rate

11

DOWN HOLE PRODUCTION CONTROL

ASPHALTENE

PARAFINS

EMULSIONS

SCALE

Long known issues seeking improved solutions

12

PIPELINES

CORROSION INHIBITORS

SCALE CONTROL

FLOW ENHANCEMENT

ELASTOMERS

Advanced chemicals, Nano paints, thermal heating, viscosity control, monitoring

13

SUCCESS HELPS MANY

- JOBS
- SUPPLY CHAIN
  - Vehicles, Steel, Chemicals
- FEEDSTOCK
- CHEMICAL PLANTS
- MANUFACTURING
- CLEANER AIR
- CHEAP ENERGY

We need you too!

14

Todays Challenge – Environmentalizing Hydraulic Fracturing
The Future – Subsurface Manufacturing?

Jeff Daniels
The Ohio State University
School of Earth Sciences

Modified from: www.blm.gov/...presentations1...ppt/Boyd_Charles-Directional_Drilling.ppt

Technological Game Changers in Subsurface Energy Production

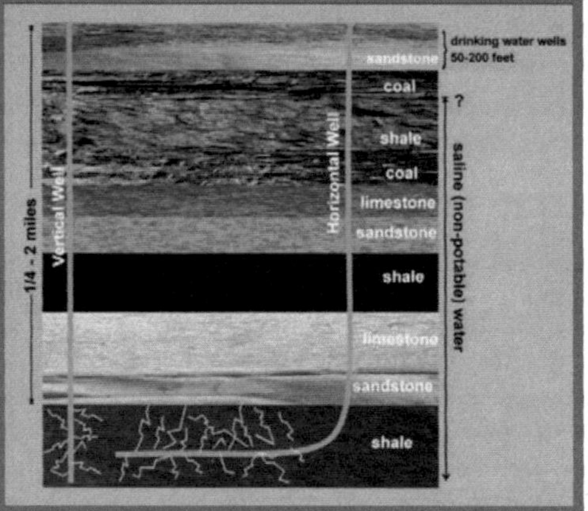

Horizontal Drilling
+
Hydraulic Fracturing
+
Gas Production from Shale

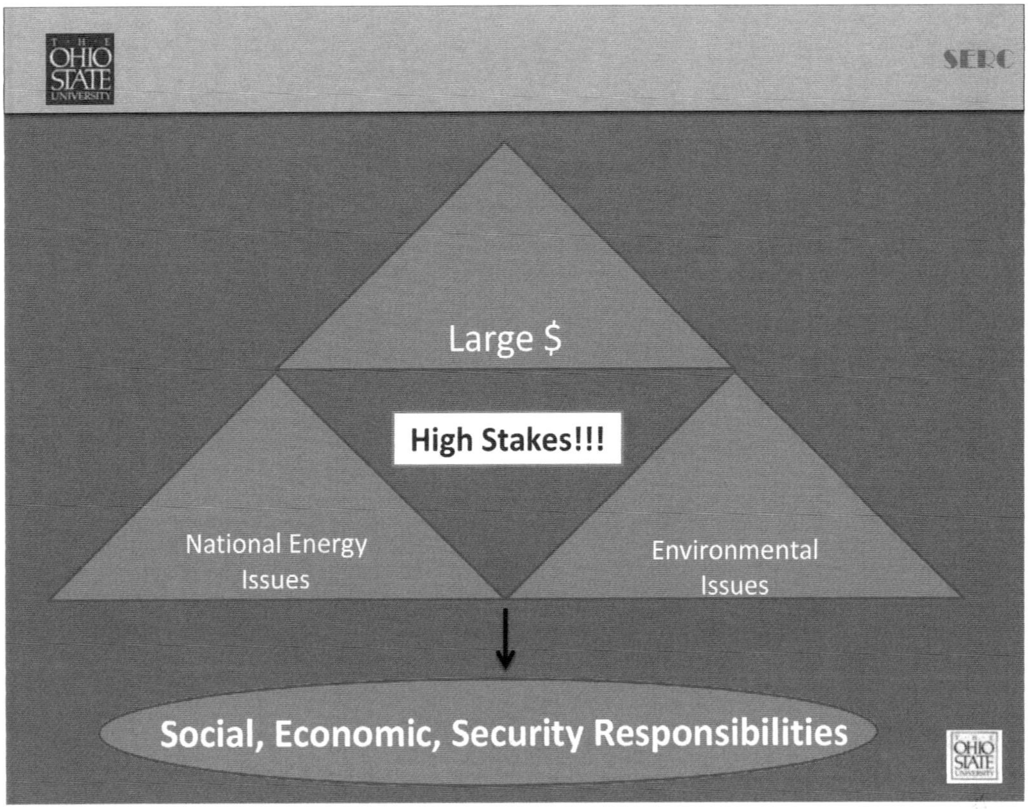

Opportunities, Challenges, and Concerns

Opportunities:

Workforce
- Direct Employment in oil and gas
- Service Company Employment
- New business development

Economic Drivers
- Billions Invested in leases
- Early Production Numbers Impressive

Challenges:

Infrastructure
- Roads, bridges, public services
- Midstream development
- Business accommodations

Environment
- Water protection
- Water sources
- Surface Impact on Wildlife

UTICA SHALE

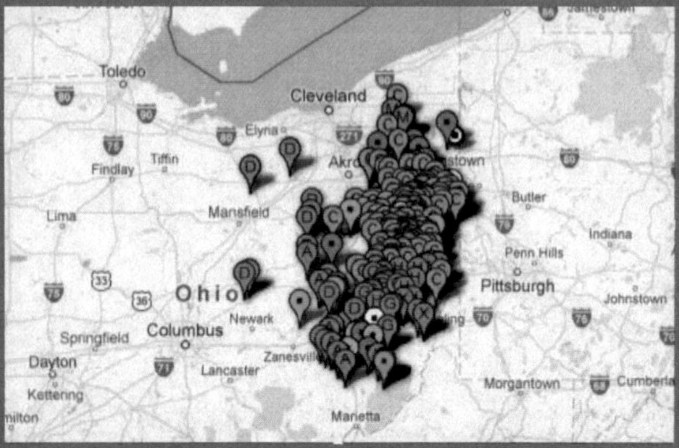

Utica Shale Activity

423 Horizontal Permits
176 Wells Drilled

UTICA SHALE

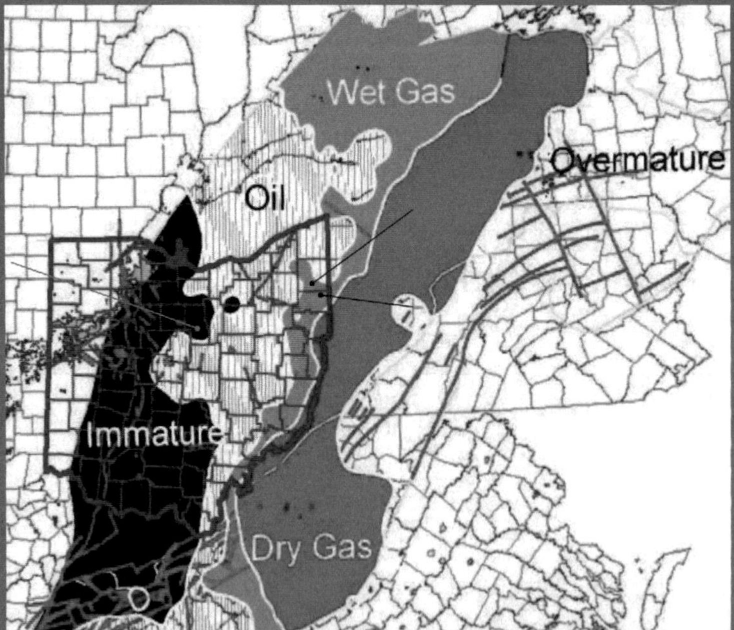

Vertical well spacing on a 40-acre grid for a 1280 acre unit would require 32 wells

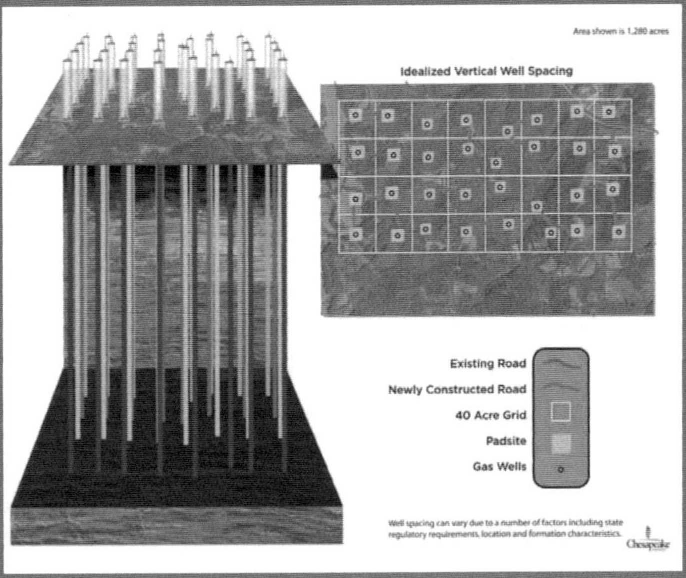

A single 5-acre pad can accommodate 8 - 12 horizontal wells to drain the same 1280-acre unit

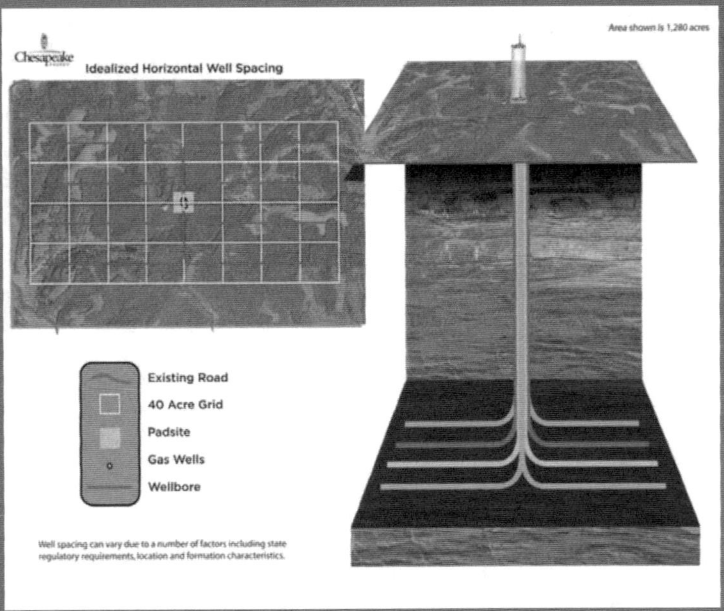

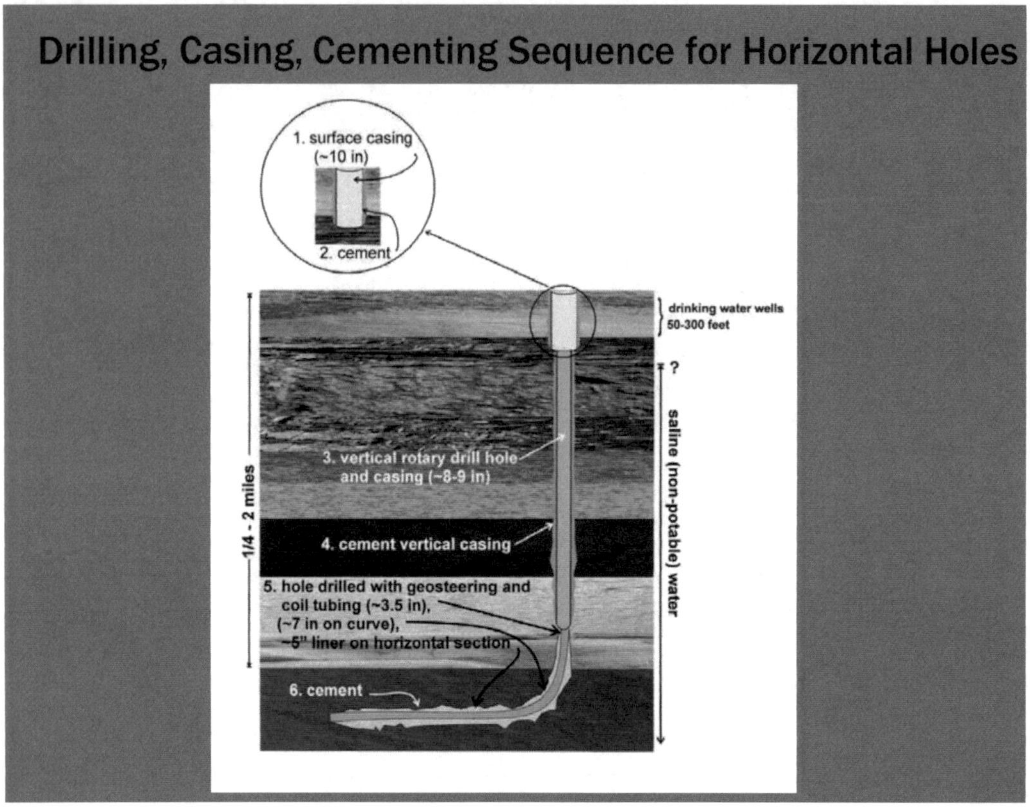
Drilling, Casing, Cementing Sequence for Horizontal Holes
1. surface casing (~10 in)
2. cement
drinking water wells 50-300 feet
?
saline (non-potable) water
1/4 - 2 miles
3. vertical rotary drill hole and casing (~8-9 in)
4. cement vertical casing
5. hole drilled with geosteering and coil tubing (~3.5 in), (~7 in on curve), ~5" liner on horizontal section
6. cement

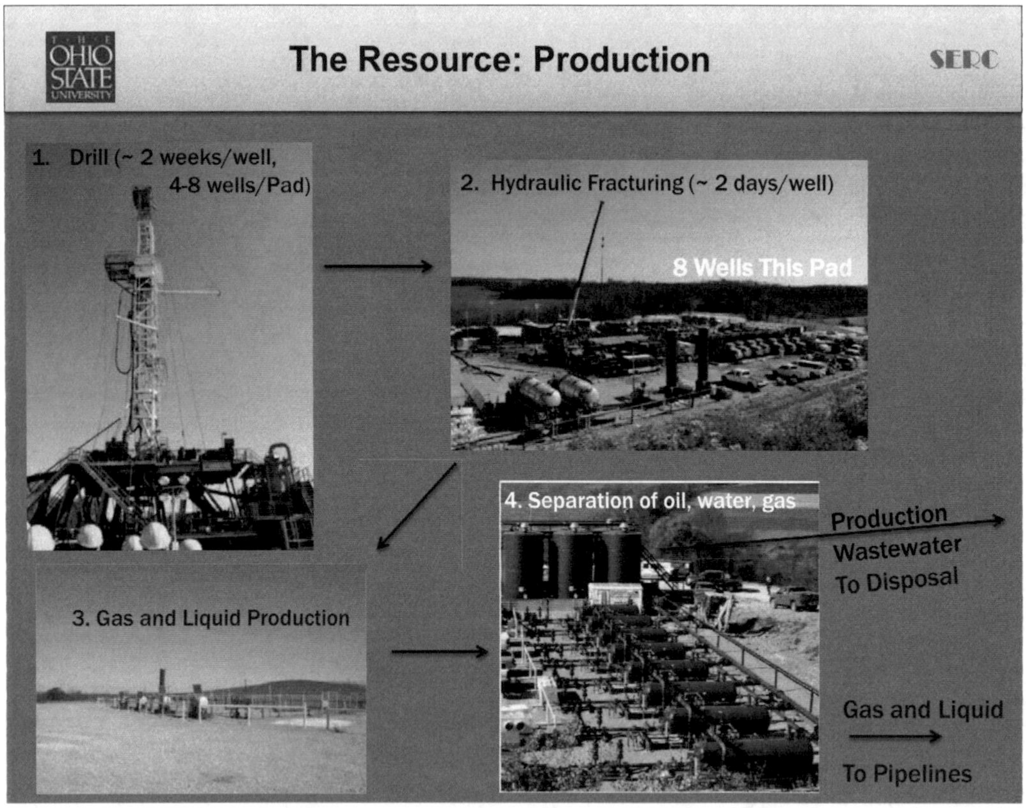
THE OHIO STATE UNIVERSITY
The Resource: Production
SERC
1. Drill (~ 2 weeks/well, 4-8 wells/Pad)
2. Hydraulic Fracturing (~ 2 days/well)
8 Wells This Pad
3. Gas and Liquid Production
4. Separation of oil, water, gas
Production Wastewater To Disposal
Gas and Liquid To Pipelines

Hydraulic Fracturing

Fluids under pressure create and force open fractures in "tight" (low perm) shale rock.

Water (>95%)
+
Proppants
+
Chemicals

Fresh water

1. Fluid under pressure causes weak zone to fracture

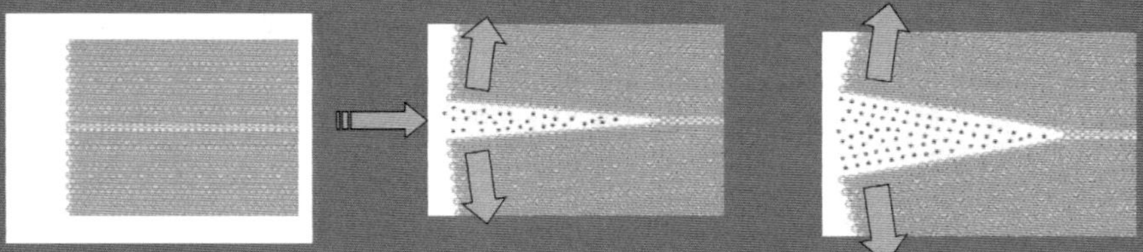

2. Proppant Particles injected into open fractures

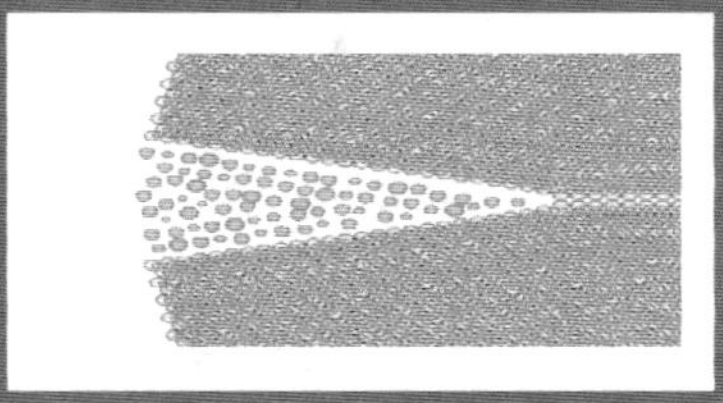

3. Natural Pressure returned Proppant Particles prop up open fractures

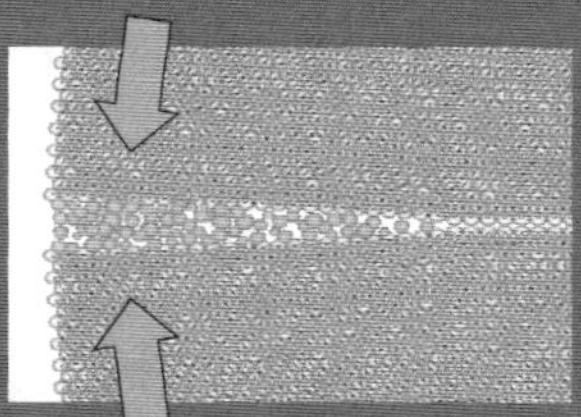

Modified from: Discontinuous Modelling of Structural Masonry and Quasi Brittle Media,
N. Bićanić, C. J. Pearce, C. Stirling, C. Davie;
NIEeS WorkshopCambridge 3-4 April 2003

Injected Hydraulic Fracturing Fluids

SERC

Proppants (Sand, or ceramic)

Fresh Water (4-6 million gallons)

Chemicals	
Biocides	Bacteria control
Clay stabilizer	Prevents swelling
Corrosion Inhibitor	Prevents pipe corrosion
Friction Reducer	"slicks" the water
Scale Inhibitor	Thickens water to suspend sand
Surfactant	Increases viscosity

Flowback Waste Water Removed

Frack Water Sources

surface water,
groundwater,
municipal water suppliers,
treated wastewater from municipal and industrial treatment facilities,
power plant cooling water, and/or
recycled produced water and/or flow back water

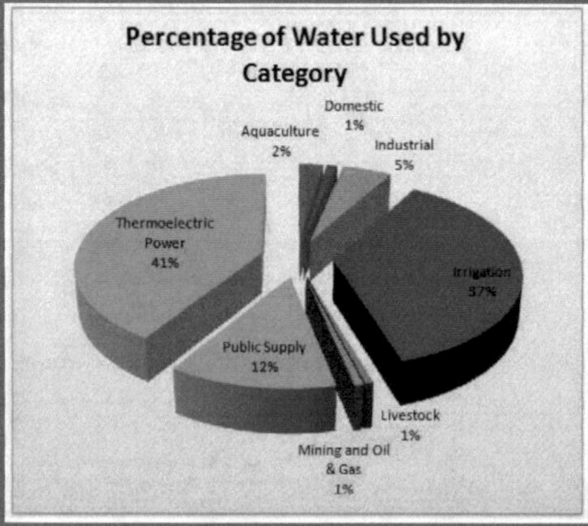

http://fracfocus.org/

Frack Fluid Chemicals

(EPA 816-R-04-003)

Table 4-1: Characteristics of Undiluted Chemicals Found in Hydraulic Fracturing Fluids (Based on MSDSs)

Product	Chemical Composition Information[1]	Hazards Information	Toxicological Information[2]	Ecological Information
Linear gel delivery system	1) 30-60% by wt. Guar gum derivative 2) 60-100% by wt. Diesel	• Harmful if swallowed • Combustible	• Chronic effects/Carcinogenicity – contains diesel, a petroleum distillate and known carcinogen • Causes eye, skin, respiratory irritation • Can cause skin disorders • Can be fatal if ingested	Slowly biodegradable
Water gelling agent	1) 60-100% by wt. Guar gum 2) 5-10% by wt. Water 3) 0.5-1.5% by wt. Fumaric acid	None	May be mildly irritating to eyes	Biodegradable
Linear gel polymer	1) <2% by wt. Fumaric acid 2) <2% by wt. Adipic acid	Flammable vapors	Can cause eye, skin and respiratory tract irritation	Not determined
Linear gel polymer slurry	1) 30-60% by wt. Diesel oil #2	• Causes irritation if swallowed • Flammable	• Carcinogenicity – Possible cancer hazard based on animal data; diesel is listed as a category 3 carcinogen in EC Annex I • May cause pain, redness, dermatitis	Partially biodegradable
Crosslinker	1) 10-30% by wt. Boric Acid 2) 10-30% by wt. Ethylene Glycol 3) 10-30% by wt. Monoethanolamine	• Harmful if swallowed • Combustible	• Chronic effects/Carcinogenicity D5 may cause liver, heart, brain reproductive system and kidney damage, birth defects (embryo and fetus toxicity) • Causes eye, skin, respiratory irritation • Can cause skin disorders and eye ailments	Not determined
Crosslinker	1) 10-30% by wt. Sodium tetraborate decahydrate	May be mildly irritating: • to eyes and skin • if swallowed	May be mildly irritating	• Partially biodegradable • Low fish toxicity
Foaming agent	1) 10-30% by wt. Isopropanol 2) 10-30% by wt. Salt of alkyl amines 3) 1-5% by wt. Diethanolamine	• Harmful if swallowed • Highly flammable	• Chronic effects/Carcinogenicity – may cause liver and kidney effects • Causes eye, skin, respiratory irritation • Can cause skin disorders and eye ailments	Not determined
Foaming agent	1) 10-30% by wt. Ethanol 2) 10-30% by wt. 2-Butoxyethanol 3) 25-55% by wt. Ester salt 4) 0.1-1% by wt. Polyglycol ether 5) 10-30% by wt. Water	Harmful if swallowed or absorbed through skin	• May cause nausea, headache, narcosis • May be mildly irritating	Harmful to aquatic organisms
Acid treatment - hydrochloric acid	1) 30-60% by wt. Hydrochloric acid	• May cause eye, skin and respiratory burns • Harmful if swallowed	• Chronic effects/Carcinogenicity – prolonged exposure can cause erosion of teeth • Causes severe burns, and skin disorders	Not determined
Acid treatment - formic acid	1) 85% by wt. Formic acid	• May cause mouth, throat, stomach, skin and respiratory tract burns • May cause genetic changes	• May cause heritable genetic damage in humans • Causes severe burns • Causes tissue damage	Not determined
Breaker Fluid	1) 60-100% by wt. Diammonium peroxidisulphate	• May cause respiratory tract, eye or skin irritation • Harmful if swallowed	May cause redness, discomfort, pain, coughing, dermatitis	Not determined

Fracking Chemicals (cont.)

Table 4-1: Characteristics of Undiluted Chemicals Found in Hydraulic Fracturing Fluids (Based on MSDSs)

Product	Chemical Composition Information[1]	Hazards Information	Toxicological Information[2]	Ecological Information
Microbicide	1) 60-100% by wt. 2-Bromo-2 nitrol,3-propanedol	• May cause eye and skin irritation	• Chronic effects/Carcinogenicity – not determined • Can cause permanent eye damage, skin disorders, abdominal pain, nausea, and diarrhea if ingested	Not determined
Biocide	1) 60-100% by wt. 2,2-Dibromo-3-nitrilopropionamide 2) 1-5% by wt. 2-Bromo-3-nitrilopropionamide	• Causes severe burns • Harmful if swallowed • May cause skin irritation; may cause allergic reaction upon repeated skin exposure	• Harmful if swallowed; large amounts may cause illness • Irritant; may cause pain or discomfort to mouth, throat, stomach; may cause pain, redness, dermatitis	Not determined
Acid corrosion inhibitor	1) 30-60% by wt. Methanol 2) 5-10% by wt. Propargyl alcohol	• May cause eye and skin irritation, headache, dizziness, blindness and central nervous system effects • May be fatal if swallowed • Flammable	• Chronic effects/Carcinogenicity – may cause eye, blood, lung, liver, kidney, heart, central nervous system and spleen damage • Causes severe eye, skin, respiratory irritation • Can cause skin disorders	Not determined
Acid corrosion inhibitor	1) 30-60% by wt. Pyridinium, 1-(Phenylmethyl)-, Ethyl methyl derivatives, Chlorides 2) 15% by wt. Thiourea 3) 5-10% Propan-2-ol 4) 1-5% Poly(oxy-1,2-ethanediyl)-nonylphenyl-hydroxy 5) 10-30% Water	• Cancer hazard (risk depends on duration and level of exposure) • Causes severe burns to respiratory tract, eyes, skin • Harmful if swallowed or absorbed through skin	• Carcinogenicity – Thiourea is known to cause cancer in animals, and possibly causes cancer in humans • Corrosive - short exposure can injure lungs, throat, and mucous membranes; can cause burns, pain, redness swelling and tissue damage	• Toxic to aquatic organisms • Partially biodegradable

[1] Information presented is for the pure product, which is significantly diluted prior to injection. MSDS chemical composition percentages may total more than 100%.
[2] Toxicity is concentration dependent.

See also:http://fracfocus.org/chemical-use/what-chemicals-are-used

Wastewater Disposal

Classification of Wells for Hydraulic Fracturing and Waste Water Disposal

Class I

Class II

Class III

Class II Wells

- Inject fluids associated with oil and natural gas production
- Most Fluid is salt water from oil and gas production
- Also may contain frack fluids
- Most Class II wells are in Texas, California, Oklahoma, and Kansas

    - Enhanced Recovery Wells
    - Disposal wells
    - Hydrocarbon storage wells

Note: Ohio is a primacy state for Class II wells.

CLASS II SALTWATER INJECTION

- Approximately 98% of oilfield fluids in Ohio are injected
- Remaining 2% is spread legally for dust and ice control
- In 2010, over 8,500,000 barrels (42 gallons per barrel) were injected
- Currently, 187 permitted Class II saltwater injection wells

From, **DEVELOPMENT OF THE SHALE PLAYS AND DISPOSAL OF OILFIELD WASTES IN OHIO**
Tom Tomastik, Geologist, Ohio DNR, Division of Oil and Gas Resources Management

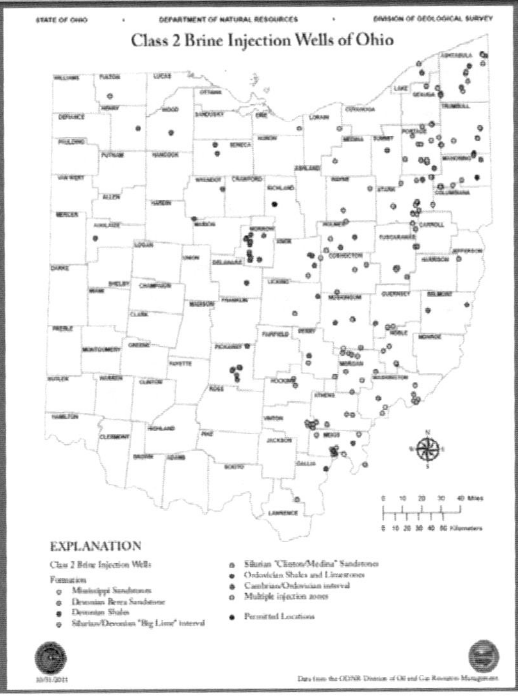

Ohio Injection Well

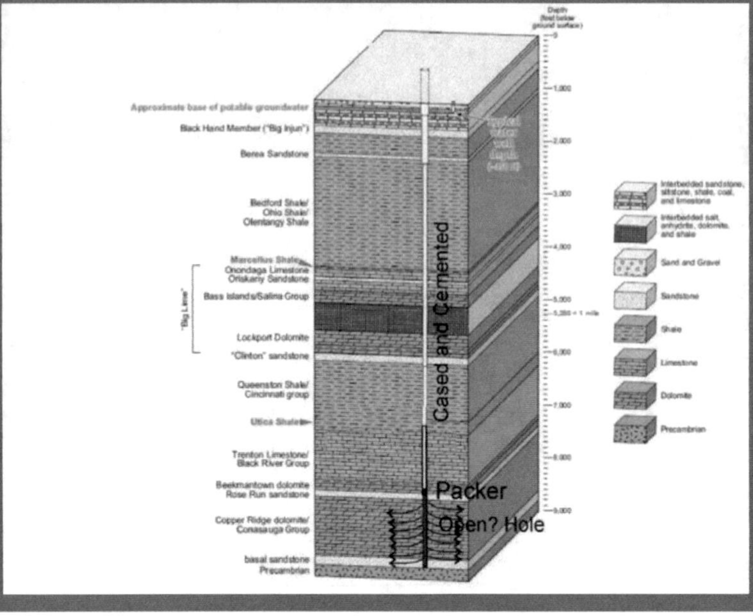

CLASS II SALTWATER INJECTION WELLS IN OHIO

- Requires three layers of steel casing to protect aquifers
- Surface casing set at least 50 feet below the deepest USDW
- Protect up to 10,000 mg/L total dissolved solids

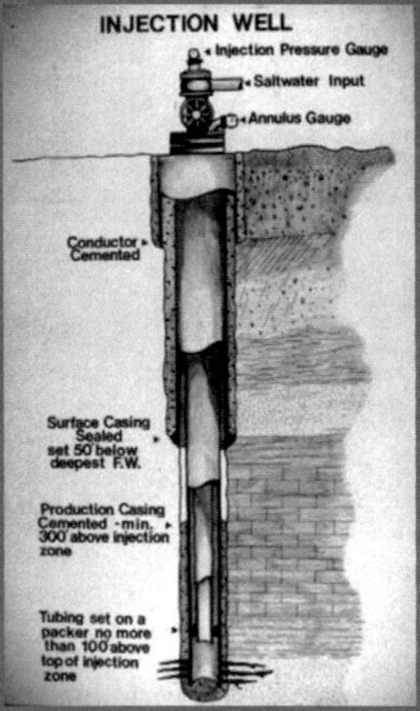

From, **DEVELOPMENT OF THE SHALE PLAYS AND DISPOSAL OF OILFIELD WASTES IN OHIO**
Tom Tomastik, Geologist, Ohio DNR, Division of Oil and Gas Resources Management

Will Shale Energy Lead to Subsurface Manufacturing?

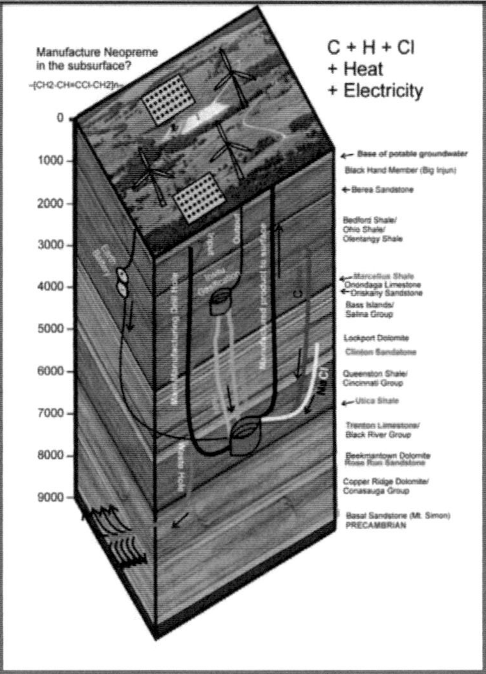

Path Forward: Learn & Cooperate

MCAFEE&TAFT

Water Law and Policy: Implications for Water Sourcing in Shale Development

AIChE and AWMA Joint Workshop
November 1, 2012

Mary Ellen Ternes
McAfee & Taft
Oklahoma City, Oklahoma

Leaders In Creating Legal Solutions®

MCAFEE&TAFT

Mary Ellen Ternes

Mary Ellen Ternes is shareholder with McAfee & Taft and a Fellow of the American College of Environmental Lawyers.

Mary Ellen serves in numerous leadership positions with AIChE and the American Bar Association. Currently, Ms. Ternes serves AIChE as Chair of the Public Affairs and Information Committee (PAIC).

Prior to entering law school, she worked as a chemical engineer for the U.S. EPA and a regulatory project officer for a commercial hazardous waste incineration company. During law school, Mary Ellen clerked by the EPA Office of General Counsel, Air and Radiation Division. In her broad environmental law practice, Mary Ellen assists many upstream oil and gas clients with air, water and waste issues, including specific issues arising from hydraulic fracturing.

Leaders In Creating Legal Solutions®

McAFEE&TAFT

Eastern Water Law "Riparian rights"

- Landowners adjacent to stream have priority
- Allows reasonable use of water flowing on land
- No unreasonable interference with other riparian owners
- "On tract" use

Leaders In Creating Legal Solutions®

McAFEE&TAFT

Western Water Law "Prior Appropriation"

- First in time, first in right
- Allows beneficial use
- May not pollute (waste) water and thus interfere with downstream users

Leaders In Creating Legal Solutions®

McAFEE&TAFT

Common Law - Groundwater

- Property of landowner which can be leased or sold
- Use for legitimate purpose
- No waste
- Compare
  - Resource to be conserved versus
  - Commodity to be used

Leaders In Creating Legal Solutions®

McAFEE&TAFT

Issues Arising Under Common Law

- Water rights are uncertain due to case-by-case adjudication
- Uncertainty impacts long-range planning and development
- No opportunity to review all potential uses simultaneously

Leaders In Creating Legal Solutions®

McAFEE&TAFT

Issues Arising Under Common Law

- Managing groundwater and surface water separately is inefficient and
- No express recognition of instream flow protection

Leaders In Creating Legal Solutions®

McAFEE&TAFT

Statutory Relief

Leaders In Creating Legal Solutions®

McAFEE&TAFT

Pennsylvania Statutory Structure

- Pennsylvania Water Rights Act of 1939
  - Surface water withdrawals for public water supply – permits
- Pennsylvania Water Resources Planning Act (Act 220) of 2002
  - Reports of withdrawals and usage by large users

Leaders In Creating Legal Solutions®

McAFEE&TAFT

Not Enough

Leaders In Creating Legal Solutions®

McAFEE&TAFT

River Basin Commissions

- Comprehensive Planning – cooperative, long-range
- Manage water by hydrologic boundaries – common standards
- Regulated riparianism – permits for withdrawals and diversions

Leaders In Creating Legal Solutions®

Delaware River Watershed Facts

- Nearly 15 million people (about 5% of the U.S. population) rely on the waters of the basin
- Drains 13,539 sq. miles, or 0.4 of 1% of the total continental U.S. land area
- 216 tributary streams
- 4 states
- 42 counties
- 838 municipalities

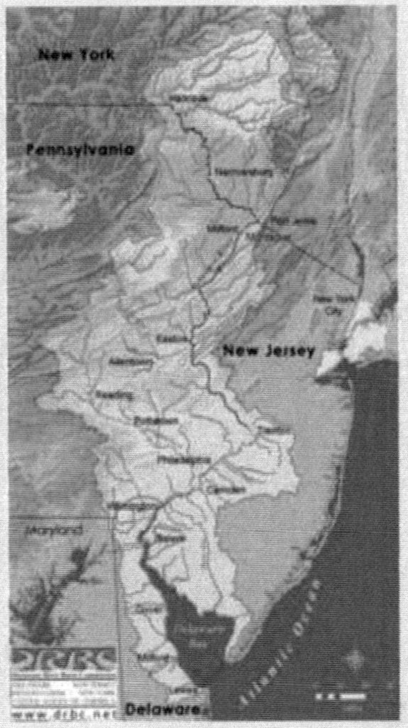

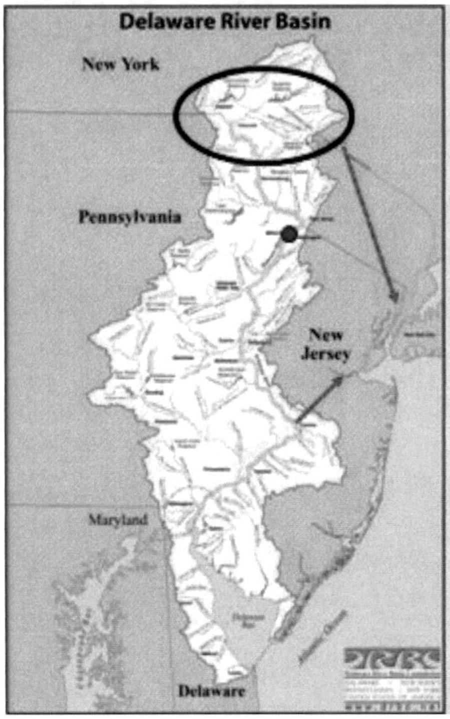

1954 U.S. Supreme Court Decree

- NYC Reservoirs – 271 BG storage
- NYC – 800 mgd withdrawal
  - (650 mgd average)
- New Jersey – 100 mgd withdrawal
- Montague flow target –
  - 1,750 cfs

- 1960s Drought –
  - Not Enough Water!

Delaware River Basin Commission

- ❑ Founded in 1961
- ❑ Five Members:
  - ➢Delaware
  - ➢New Jersey
  - ➢Pennsylvania
  - ➢New York State
  - ➢Federal Government

McAFEE&TAFT

Powers of the DRBC

- Comprehensive Planning
  - Projects
- Adjudication
  - Project approvals for any federal, state, local or private project having a substantial effect on waters of the basin
    - Includes water withdrawal approvals
- Regulation
  - Includes operation of reservoirs during drought conditions, water quality objectives

Leaders In Creating Legal Solutions®

McAFEE&TAFT

Groundwater Protected Areas

- Withdrawal permits to avoid depletion of natural stream flows and groundwater
- Regulations – withdrawal limits to protect reliability of water suppliers, and prevent adverse impact on water quality, storage capacity or low flows of perennial streams
- Conjunctive use (storing surface water in a groundwater basin in wet years, withdrawing it from the basin in dry years)
- Water Conservation
- New users – mitigation measures

Leaders In Creating Legal Solutions®

McAFEE&TAFT

Special Protection Waters and Regulation

- Antidegradation program for areas with exceptionally high scenic, recreational, ecological and/or water supply values
- Regulation requires:
  - No measurable change to existing water quality at control points
  - Cumulative impact analysis
  - Nonpoint source pollution control plan

Leaders In Creating Legal Solutions®

McAFEE&TAFT

Water Demand in Marcellus Shale

- Vertical well: 85,000 gallons for drilling
- Horizontal well with hydraulic fracturing: 5 million gallons per stimulation
- Use of water for well stimulation is small in comparison to other uses
- Estimated future use of water in Delaware River Basin for natural gas exploration and production – **0.24% of total withdrawals**

Leaders In Creating Legal Solutions®

McAFEE&TAFT

Sheer volume versus Location and Timing of Withdrawal

Leaders In Creating Legal Solutions®

McAFEE&TAFT

DRBC Response to Shale Gas Projects

- 2009 - Sponsors of natural gas projects in shale formations in drainage areas of Special Protection Waters required to obtain commission approval
- 2010 - Exploratory wells without prior state approval required to have commission approval
- 2010 – Natural Gas Regulations proposed
- 2011 – DRBC postpones DRBC review of any applications for water withdrawals within NY intended for high volume hydrofracking
- 2011 – Natural Gas Regulations revised, not yet adopted

Leaders In Creating Legal Solutions®

McAFEE&TAFT

Revised Draft DRBC Regulations

- Water sources and conditions of "bulk water use" and management approval of natural gas development projects
- Natural gas development plans to protect high value water resource landscapes and Special Protection Waters
- Wastewater shipments tracked and wastewater regulated to prevent pollutant discharges
- DRBC would rely on state requirements to avoid duplicative regulation and defer to more stringent requirements

Leaders In Creating Legal Solutions®

McAFEE&TAFT

Implications for Water Sourcing

- Managed water based on hydrologic boundaries
- Comprehensive planning – long range, balancing all present and future uses
- Coordinate programs of all government agencies
- Mitigation, monitoring and reporting
- Conservation, recycling

Leaders In Creating Legal Solutions®

McAFEE&TAFT

Implications for Water Sourcing

- Cumulative impact analysis
- Permits to regulate quantity, timing and manner of withdrawal, conjunctive use
- Instream flow protection
- Sound science, adaptive management
- Innovation

Leaders In Creating Legal Solutions®

McAFEE&TAFT

References

- Kenneth J. Warren, "Water Supply for Shale Gas Production: Lessons from the River Basin Commission Management in the Mid-Atlantic States," 58 Rocky Mt. Min. L. Inst. 9-1 (2012). Mr. Warren serves as general counsel to the Delaware River Basin Commission.
- Delaware River Basin Commission, DRBC's Water Quality Regulations, Article 7, available at: http://www.state.nj.us/drbc/programs/natural/draft-regulations.html
- DRBC, "Natural Gas Development Revised Draft Regulation" (Nov. 8, 2011) available at: http://www.state.nj.us/drbc/library/documents/naturalgas-REVISEDdraftregs110811.pdf.

Leaders In Creating Legal Solutions®

Regulating Water Sources for Shale Gas Development in the Susquehanna River Basin

Shale Oil and Gas E&P - Water Challenges and Opportunities
AIChE-A&WMA Workshop November 1, 2012

Susquehanna River Basin Commission www.srbc.net

Overview

- Background
- Project Review & Approval
- Regulatory Changes Related to Natural Gas
- Low Flow Protection Requirements
- Water Use Profile
- Website Resources

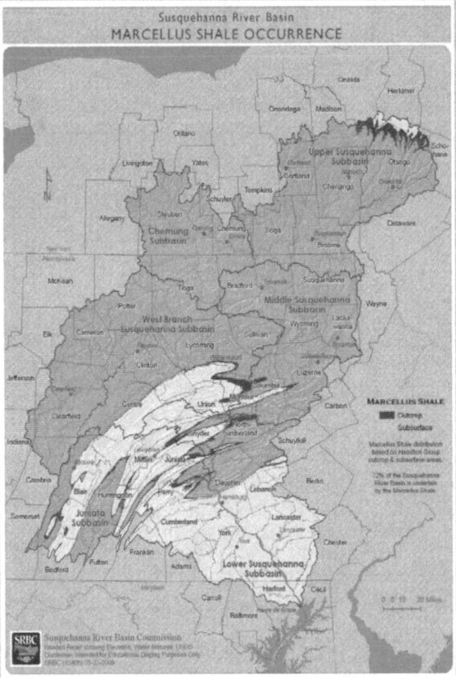

Susquehanna River Basin Commission www.srbc.net

Susquehanna River Basin

The Basin

- 27,510-square-mile watershed
- Comprises 43 percent of the
- Chesapeake Bay watershed
- 4.2 million population
- 60 percent forested
- 32,000+ miles of waterways

The Susquehanna River

- 444 miles, largest tributary to the Chesapeake Bay
- Supplies 18 million gallons a minute to the Bay

Susquehanna River Basin Commission www.srbc.net

Commission Background

- Federal-Interstate Compact Commission
- New York, Pennsylvania, Maryland and the United States
- Broad Water Resource Management Authority
  - Manage resource impacts of projects
  - Avoid conflicts with other users
  - Standards for equal and uniform treatment

Susquehanna River Basin Commission www.srbc.net

Commission Programs

- Flood Forecast and Warning System
- Monitoring and Assessment
- Public Education and Outreach
- Water Resources Planning
- Drought Coordination
- Regulatory Program

Susquehanna River Basin Commission www.srbc.net

What We Don't Regulate

- We don't regulate mineral resources
- We don't regulate the drilling, production or transmission of natural gas
- We don't regulate the treatment or disposal of flowback and production fluids
- We don't regulate water quality
- Specifically, we rely on member jurisdictions for standards for well casing, treatment and disposal, and avoid duplication in areas like water quality

Susquehanna River Basin Commission www.srbc.r

What We Do Regulate

- We regulate all water withdrawals associated with unconventional hydrocarbon development
- We regulate all water sources associated with unconventional hydrocarbon development
- We regulate consumptive use of water associated with unconventional hydrocarbon development
- Specifically, we do regulate is water resources and their use associated with natural gas development

Susquehanna River Basin Commission www.srbc.n

General Regulatory Authorities

•Groundwater withdrawals
  100,000 gpd (30-day average)
•Surface water withdrawals
  100,000 gpd (30-day average)
•Consumptive Use
  20,000 gpd (30-day average)
•Diversions
  In-basin: any amount
  Out-of-basin: 20,000 gpd (30-day average)
•Natural Gas
  Withdrawals and Consumptive Use (1st gallon)

Susquehanna River Basin Commission www.srbc.n

Project Review Applications
Requirements and Review Standards

- Notification Requirements
- Agency Coordination
- Water Source Evaluation & Testing
- Environmental Review
- Safe Yield Calculation
- Cumulative Water Use Assessment
- Low Flow Protection

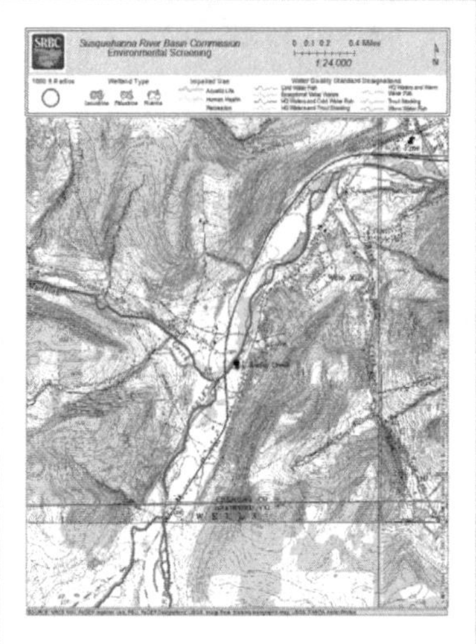

Susquehanna River Basin Commission www.srbc.

Groundwater Withdrawals

Groundwater availability analysis

- Recharge area delineation / recharge rates
- 1-in-10 year drought recurrence interval

Constant rate aquifer testing requirement

- Prior review and approval
- 72-hour testing / pre and post monitoring

Susquehanna River Basin Commission www.srbc.

Surface Water Withdrawals

- Waterbody classification
- Aquatic resource survey
- Aquatic invasive species
- Intake design & metering

Susquehanna River Basin Commission www.srbc.n

Consumptive Use

- Cumulative impact
- Sustain flows to Chesapeake Bay
- Mitigation standard: 1 – 1
- Mitigation options
- Low flow augmentation

Susquehanna River Basin Commission www.srbc.n

Project Review Approvals

- Metering
- Monitoring and Reporting
- Mitigation for Adverse Impacts
- Water Conservation
- Low Flow Protection / Passby Flow
- Compliance Inspections
- Docket Reopener

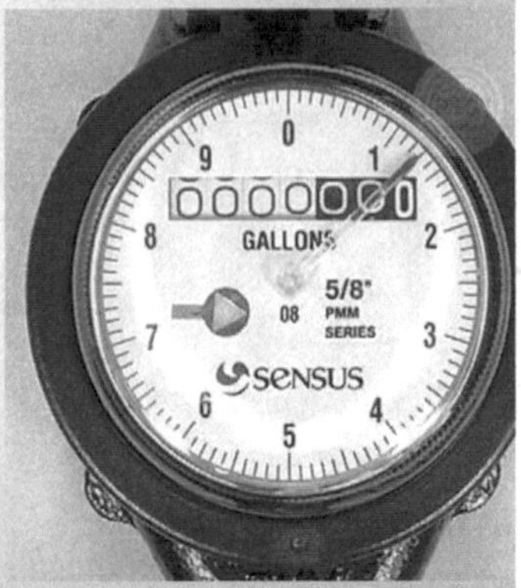

Susquehanna River Basin Commission www.srbc.

Challenges Posed By Natural Gas Development

- Nomadic and short-term nature of projects,
- Operates in headwaters/ historically less developed portions of basin,
- Withdrawals distant from consumptive use locations,
- Unpredictable pace of industry,
- Number of applications,
- Public scrutiny/opposition.

Susquehanna River Basin Commission www.srbc.

Natural Gas Regulatory Adjustments

- **July 2008: Approval by Rule process**
  Public water supply sources
- **August 2008: Notice of Determination**
  Elimination of thresholds/100% consumptive
- **January 2009: <u>First Rule Change</u>**
  Specific natural gas ABR process/gallon 1 threshold
  Addition of SRBC-approved sources to ABRs
  Lesser quality water incentives
  Certification of disposal in accordance w/rules
- **November 2009: <u>Second Rule Change</u>**
  Source list by company (not by ABR)
  Source flexibility/source sharing

Susquehanna River Basin Commission www.srbc.r

Natural Gas Regulatory Adjustments

- **November 2010: <u>Third Rule Change</u>**
  Standardized notification requirements
  Standardized transfer of approval process
  Clarification of administrative appeal process
- **December 2011: <u>Fourth Rule Change</u>**
  Defines new terms, i.e. hydrocarbon development project
  Approves interbasin transfers of flowback administratively
  Authorizes renewals
  Broadens the scope of ABRs
  Provides procedures for further water source approvals
- **December 2012: <u>Fifth Rule Change??</u>**
  Limitations on withdrawals in headwaters

Susquehanna River Basin Commission www.srbc.r

More than half of the withdrawals have passbys

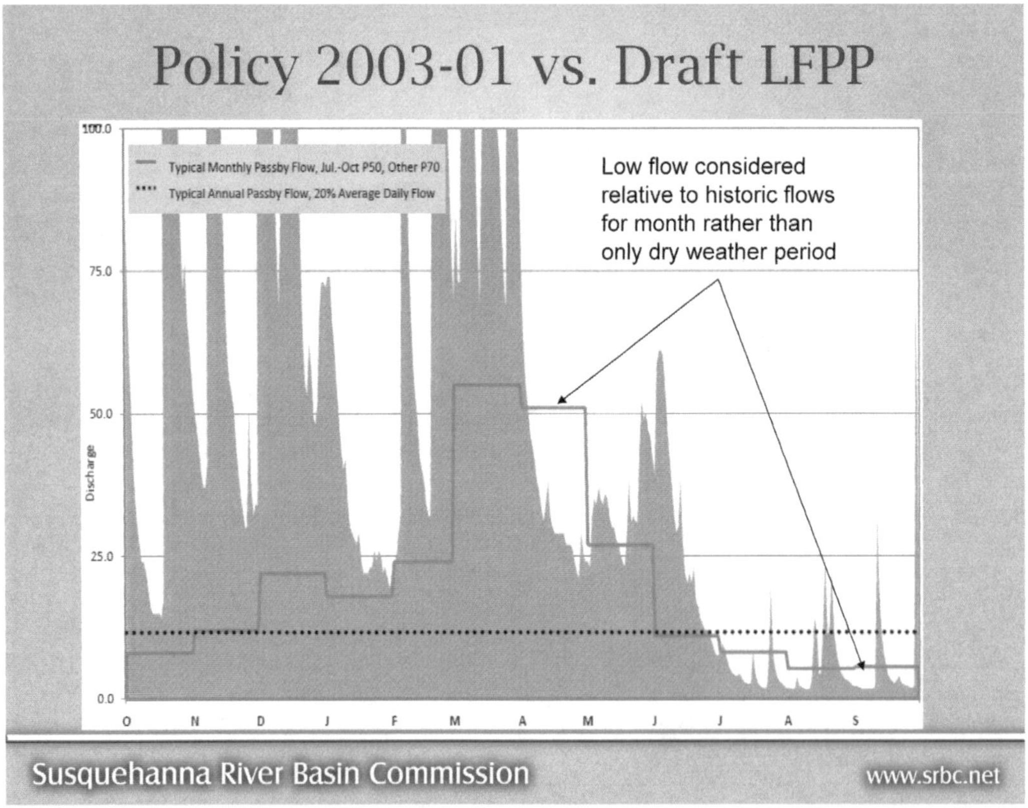
Policy 2003-01 vs. Draft LFPP
Typical Monthly Passby Flow, Jul.-Oct P50, Other P70
Typical Annual Passby Flow, 20% Average Daily Flow
Low flow considered relative to historic flows for month rather than only dry weather period
100.0
75.0
50.0
25.0
0.0
Discharge
O
N
D
J
F
M
A
M
J
J
A
S
Susquehanna River Basin Commission
www.srbc.net

Improvements Contained in the LFPP

- Establishes criteria to preserve seasonal variability of natural flow regimes.
- Establishes basinwide aquatic resource classification system
- Prescribes limits on natural flow alterations based on monthly or seasonal basis.
- Sets hierarchy of flow protection based on ecological needs and specific passby flow or conservation release values on monthly or seasonal basis.
- Improves protection of headwaters.

Susquehanna River Basin Commission www.srbc.n

Revised LFPP will be Balanced between:

- protecting the basin's water/aquatic resources and
- use of the basin's water resources.

Currently reviewing 60 comments, preparing revisions for release later this fall.

Susquehanna River Basin Commission www.srbc.net

Shale Gas In the SRB: Approvals, Water Use Profile and Trends

Susquehanna River Basin Commission www.srbc.n

Natural Gas Approvals

As of September 30, 2012

- Surface Water Withdrawals – 252
- Groundwater Withdrawals – 9
- Public Water Supply Systems – 35
- Wastewater Effluent/AMD - 2
- Approvals by Rule (Well Pads) – 2,022

Susquehanna River Basin Commission www.srbc.n

Approvals by Rule

Well Drilling Pads

Total Approved – 2,022*

- 2008 – 54
- 2009 – 321
- 2010 – 989
- 2011 – 457
- 2012 – 255*

* As of September 30, 2012

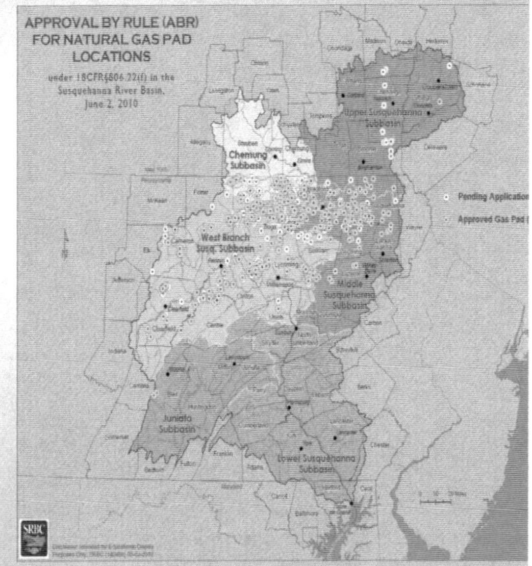

Susquehanna River Basin Commission www.srbc.

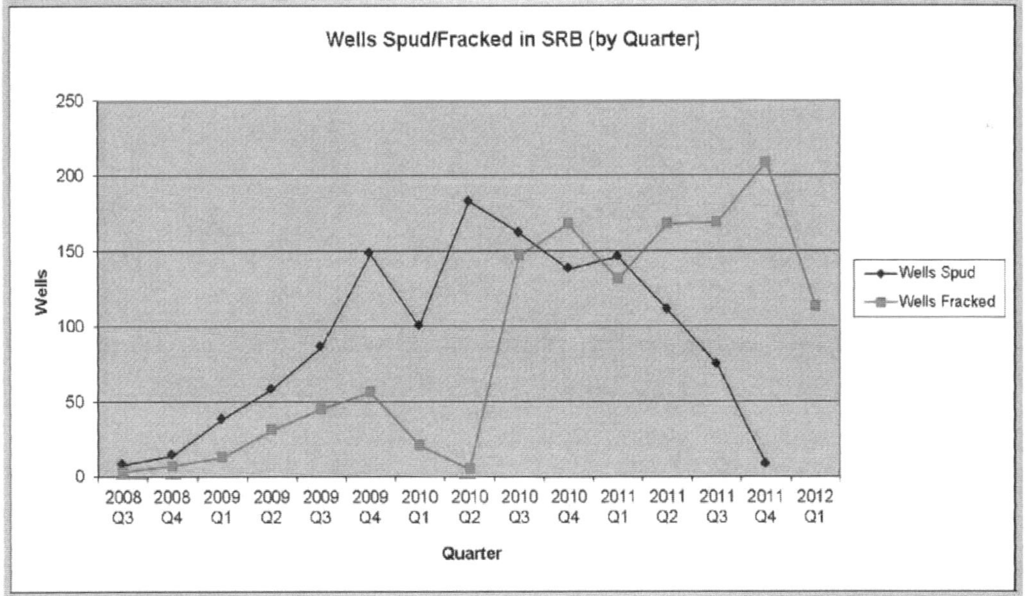

Susquehanna River Basin Commission www.srbc.net

Natural Gas Water Use Profile

(July 1, 2008 thru June 30, 2012)

- 1,594 Unconventional Gas Wells Hydro-fractured,
- 8.8 billion gallons of water consumptively used,
- 7.1 billion gallons of water withdrawn from docketed sources,
- 1.7 billion gallons of water withdrawn from non-docketed approved sources (primarily PWS),

Susquehanna River Basin Commission www.srbc.

Natural Gas Water Use Profile

continued

- 4.4 million gallons (average) used per well for hydro-fracturing (88% freshwater, 12% flowback reused).
- 8% - 10% of injected water is recovered as flowback within the first 30 days after release of pressure.
- 87% of water brought on-site is used, balance of 13% temporarily stored on site, or transferred to another site.

Susquehanna River Basin Commission www.srbc.

Where Does the Gas Industry Get Its Water?

Fresh Water Sources:

- Surface Water Withdrawals – 68%
- Groundwater Withdrawals – <1%
- Public Water Supply Systems – 32%
- Wastewater Effluent/AMD – <1%
- Water Pipelines with multiple sources – <1%

Flowback: increased from 0.430 to 0.740 mgd/well

Susquehanna River Basin Commission www.srbc.r

Current Analysis

- Current volume of water use can be accommodated,
- Timing and location of withdrawals are key,
- Interruptible and alternate sources minimize impacts during low flows,
- Science-based decision making,
- Ongoing cumulative impact assessment is critical,
- Reliance on member jurisdiction standards for well, casing, and treatment and disposal,
- Dynamic regulatory program with continued refinements.

Susquehanna River Basin Commission www.srbc.r

SRBC On-Line Resources: *www.srbc.net*

Water Resource Portal

www.srbc.net/wrp/Default.aspx

Remote Water Quality Monitoring Network

www.srbc.net/programs/remotenetwork.htm

Marcellus Shale & Natural Gas Well Development

(regulations, info. sheets, ppts, testimonies, maps, related press releases, etc.)

www.srbc.net/programs/projreviewmarcellus.htm

Susquehanna River Basin Commission www.srbc.net

Questions?

Photo Credit: TNC, 2010

Susquehanna River Basin Commission www.srbc.net

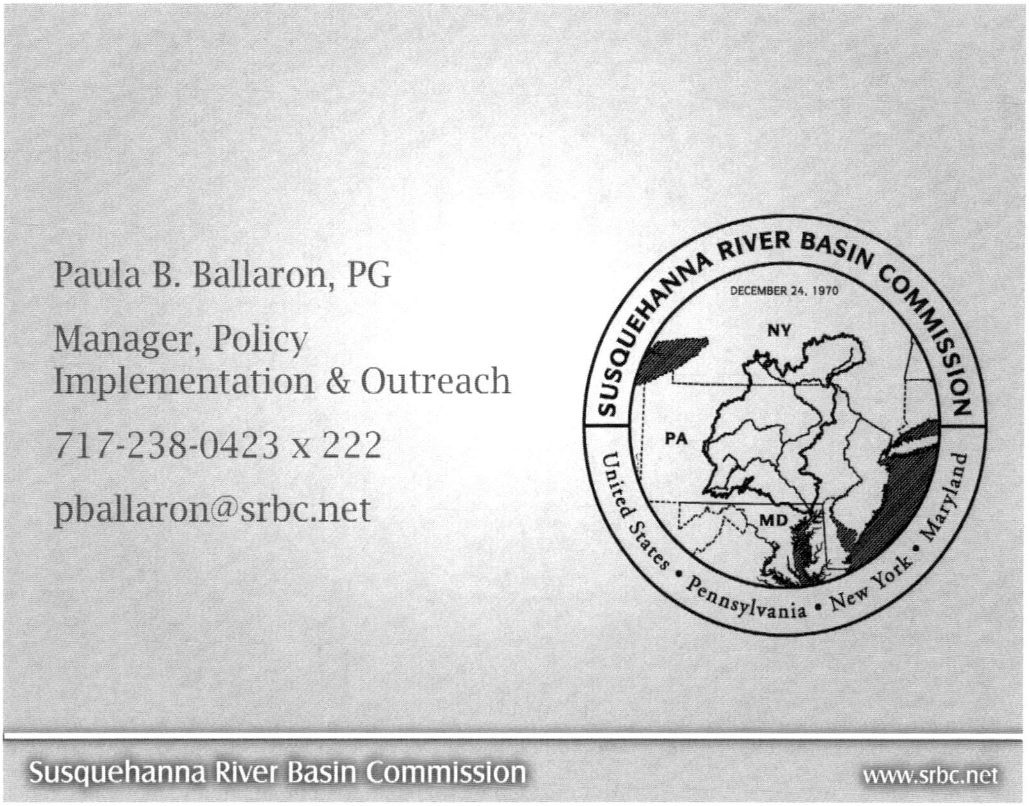
Paula B. Ballaron, PG
Manager, Policy Implementation & Outreach
717-238-0423 x 222
pballaron@srbc.net
SUSQUEHANNA RIVER BASIN COMMISSION
DECEMBER 24, 1970
NY
PA
MD
United States • Pennsylvania • New York • Maryland
Susquehanna River Basin Commission
www.srbc.net

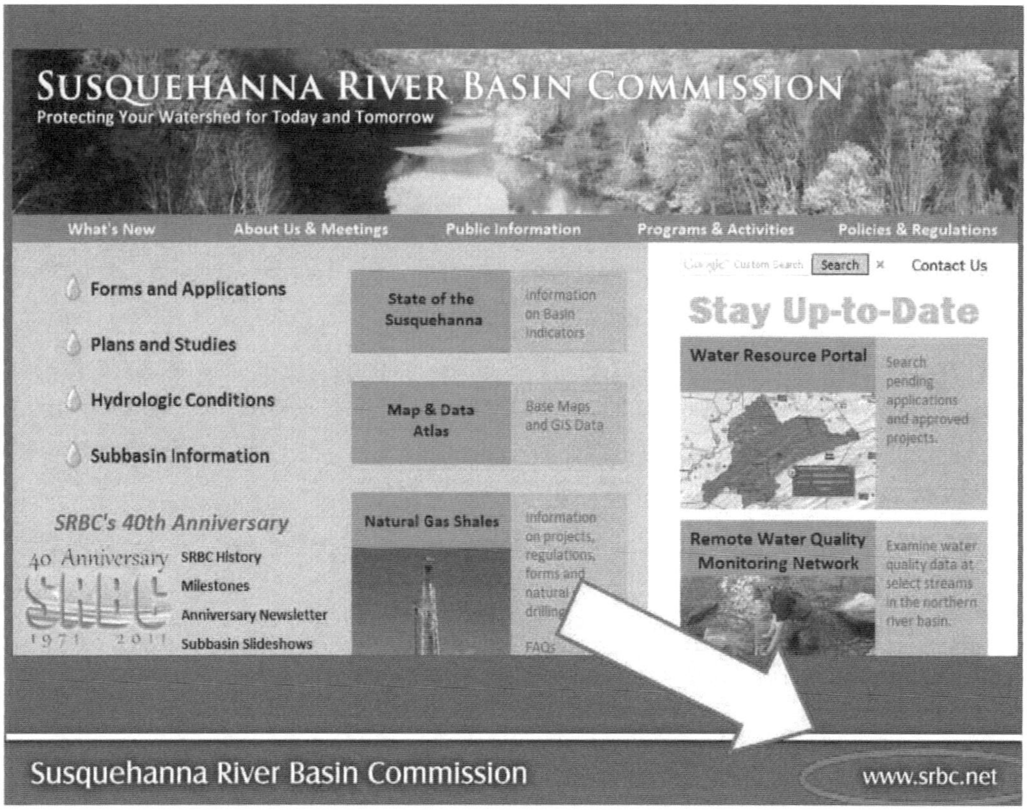
SUSQUEHANNA RIVER BASIN COMMISSION
Protecting Your Watershed for Today and Tomorrow
What's New
About Us & Meetings
Public Information
Programs & Activities
Policies & Regulations
Forms and Applications
Plans and Studies
Hydrologic Conditions
Subbasin Information
SRBC's 40th Anniversary
40 Anniversary
SRBC
1971 2011
SRBC History
Milestones
Anniversary Newsletter
Subbasin Slideshows
State of the Susquehanna
Information on Basin Indicators
Map & Data Atlas
Base Maps and GIS Data
Natural Gas Shales
FAQs
Search
Contact Us
Stay Up-to-Date
Water Resource Portal
Search pending applications and approved projects.
Remote Water Quality Monitoring Network
Examine water quality data at select streams in the northern river basin.
Susquehanna River Basin Commission
www.srbc.net

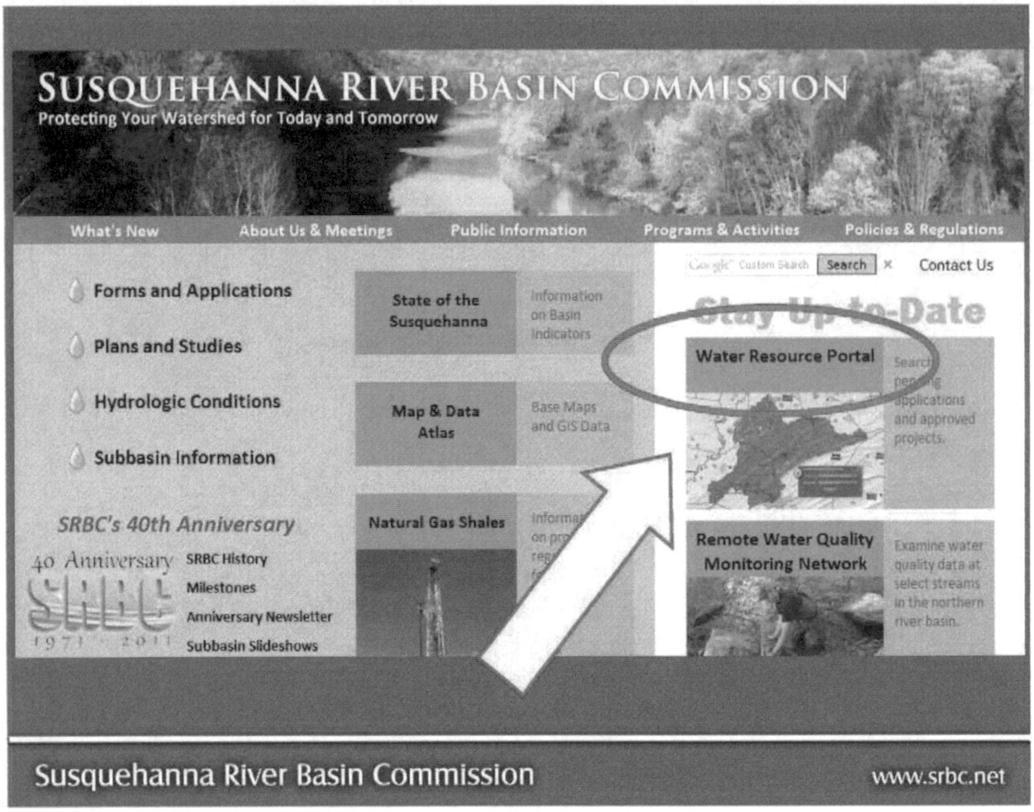
SUSQUEHANNA RIVER BASIN COMMISSION
Protecting Your Watershed for Today and Tomorrow
What's New
About Us & Meetings
Public Information
Programs & Activities
Policies & Regulations
Search
Contact Us
Forms and Applications
Plans and Studies
Hydrologic Conditions
Subbasin Information
State of the Susquehanna
Information on Basin Indicators
Map & Data Atlas
Base Maps and GIS Data
Natural Gas Shales
Stay Up-to-Date
Water Resource Portal
Remote Water Quality Monitoring Network
Examine water quality data at select streams in the northern river basin.
SRBC's 40th Anniversary
40 Anniversary
SRBC
1971 2011
SRBC History
Milestones
Anniversary Newsletter
Subbasin Slideshows
Susquehanna River Basin Commission
www.srbc.net

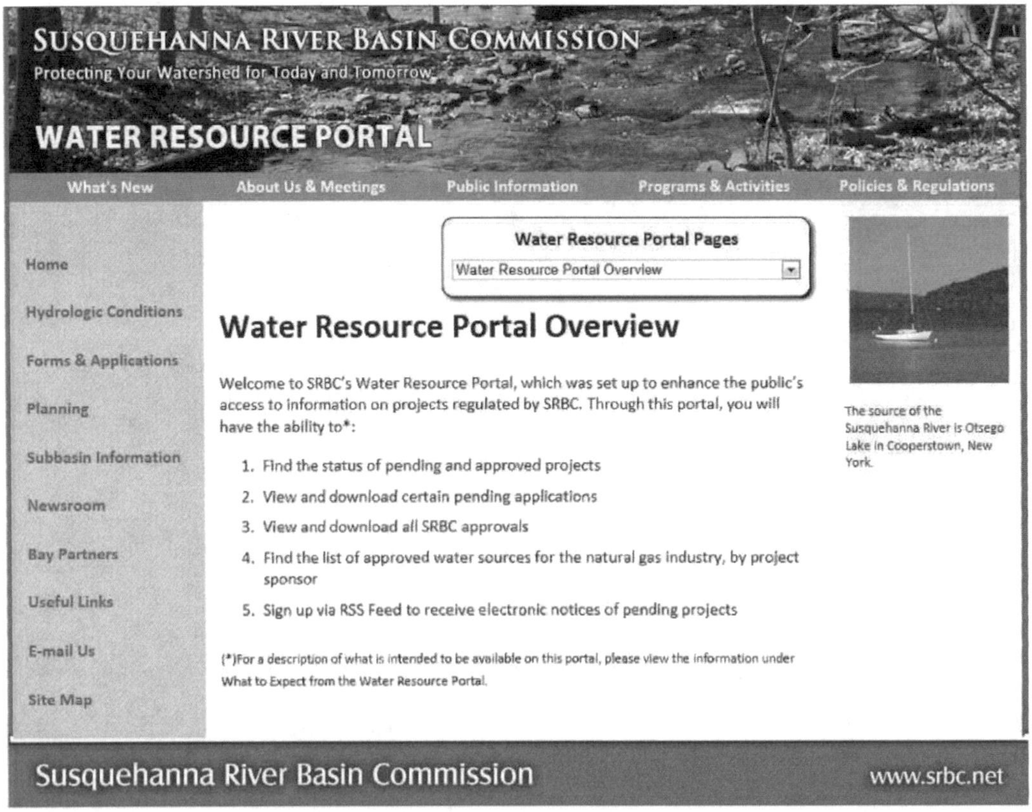
SUSQUEHANNA RIVER BASIN COMMISSION
Protecting Your Watershed for Today and Tomorrow
WATER RESOURCE PORTAL
What's New
About Us & Meetings
Public Information
Programs & Activities
Policies & Regulations
Home
Hydrologic Conditions
Forms & Applications
Planning
Subbasin Information
Newsroom
Bay Partners
Useful Links
E-mail Us
Site Map
Water Resource Portal Pages
Water Resource Portal Overview
Water Resource Portal Overview
Welcome to SRBC's Water Resource Portal, which was set up to enhance the public's access to information on projects regulated by SRBC. Through this portal, you will have the ability to*:
1. Find the status of pending and approved projects
2. View and download certain pending applications
3. View and download all SRBC approvals
4. Find the list of approved water sources for the natural gas industry, by project sponsor
5. Sign up via RSS Feed to receive electronic notices of pending projects
(*)For a description of what is intended to be available on this portal, please view the information under What to Expect from the Water Resource Portal.
The source of the Susquehanna River is Otsego Lake in Cooperstown, New York.
Susquehanna River Basin Commission
www.srbc.net

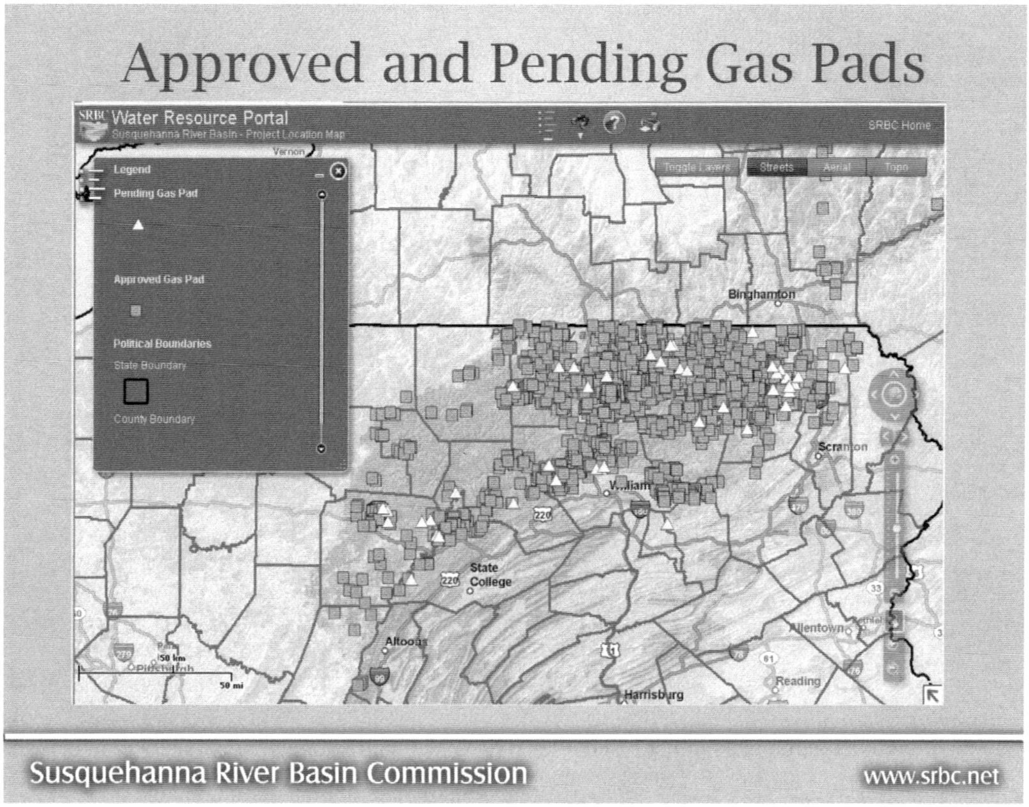
Approved and Pending Gas Pads
SRBC Water Resource Portal
Susquehanna River Basin - Project Location Map
SRBC Home
Toggle Layers
Streets
Aerial
Topo
Legend
Pending Gas Pad
Approved Gas Pad
Political Boundaries
State Boundary
County Boundary
Vernon
Binghamton
Scranton
State College
Altoona
Allentown
Reading
Harrisburg
Pittsburgh
50 km
50 mi
Susquehanna River Basin Commission
www.srbc.net

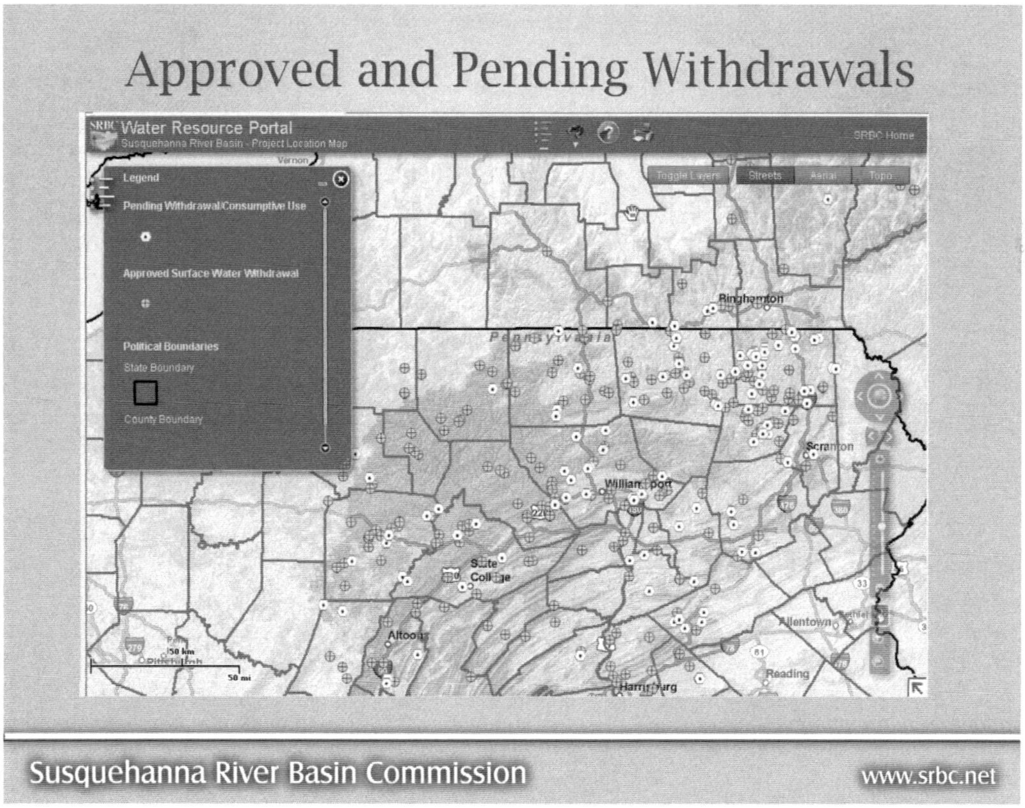
Approved and Pending Withdrawals
SRBC Water Resource Portal
Susquehanna River Basin - Project Location Map
SRBC Home
Toggle Layers
Streets
Aerial
Topo
Legend
Pending Withdrawal/Consumptive Use
Approved Surface Water Withdrawal
Political Boundaries
State Boundary
County Boundary
Vernon
Binghamton
Scranton
Altoona
Allentown
Reading
50 km
50 mi
Susquehanna River Basin Commission
www.srbc.net

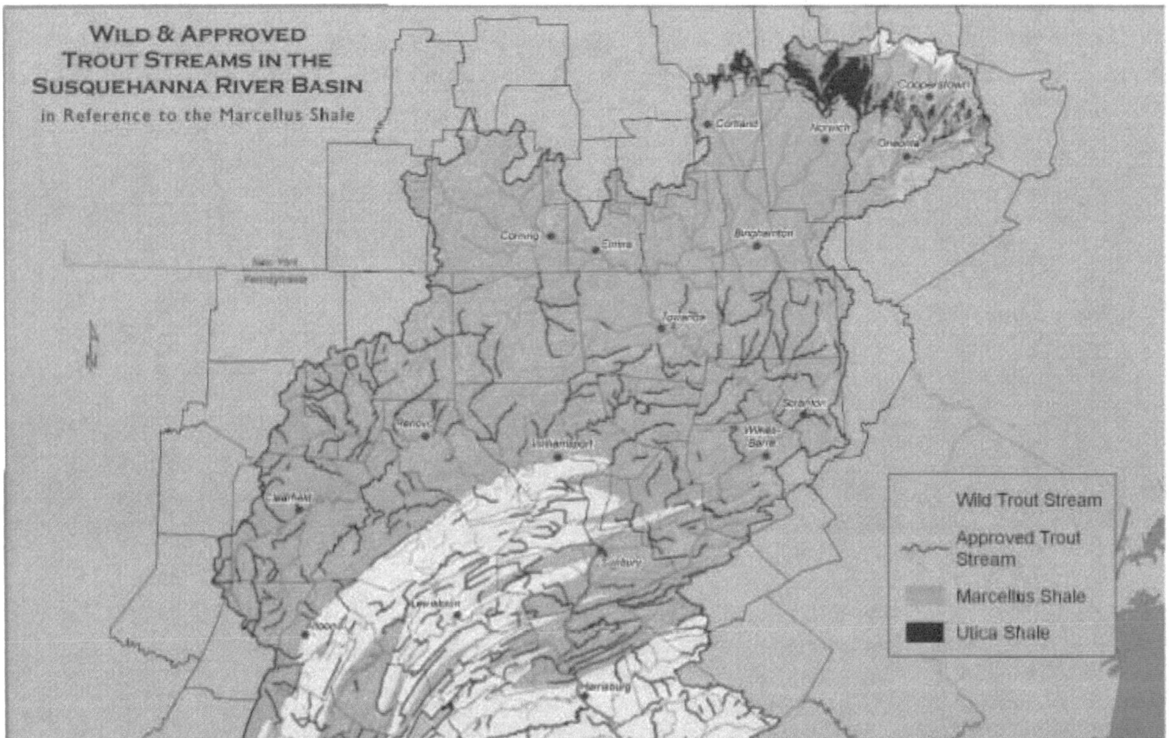
WILD & APPROVED
TROUT STREAMS IN THE
SUSQUEHANNA RIVER BASIN
in Reference to the Marcellus Shale
Wild Trout Stream
Approved Trout Stream
Marcellus Shale
Utica Shale

take a closer look

Water Resources – Alternate Water Sources / Reused / Recycled

AIChE-A&WMA Workshop on Shale Oil and Gas E&P - Water Challenges and Opportunities

Sam Bacon
Water Resources Lead
Encana Oil & Gas (USA) Inc
Mid Continent Business Unit (MCBU)

November 1, 2012

encana. natural gas

Encana Mid Continent Business Unit

encana.

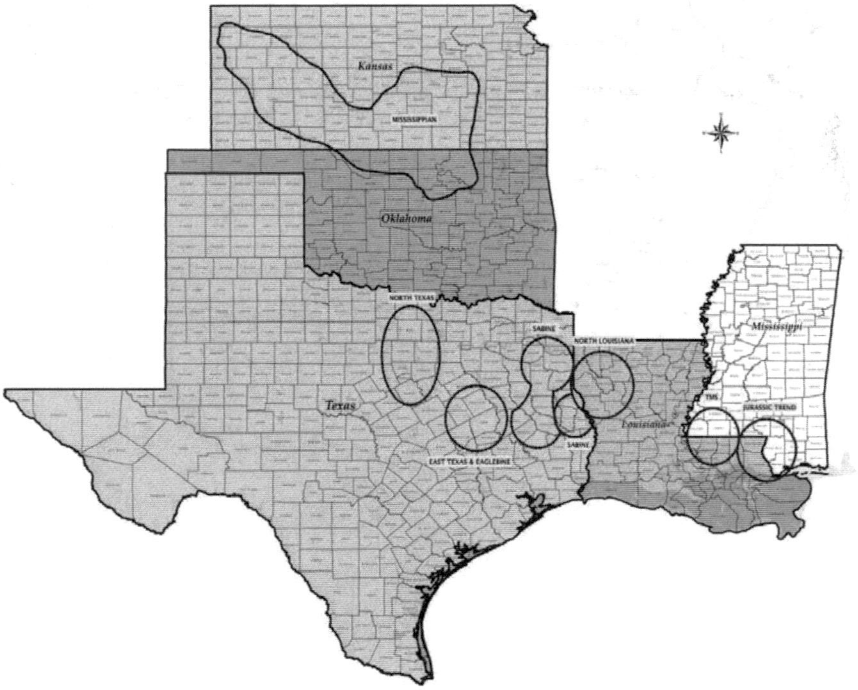

1

Water Sourcing– Mandates

- ONLY USE WATER FOR WHICH THERE IS
  - "NO SIGNIFICANT DEMAND"
    - SAVE MONEY
    - GET IT DONE

2

Water Sourcing Options – Order of Preferred Use

- Alternative Fresh Water Sources
  - POTW Effluent – Treated Waste (No Demand)
  - Alluvial Water Wells (No Significant Demand)
- Non-Fresh Water Sources
  - Brine / Brackish Water (No Significant Demand)
- Reused / Repurposed Water
  - Produced Water – (No Demand)
  - Flowback - (No Demand)
  - E&P Liquids from Drilling Operations - (No Demand)
- Fresh Water Sources
  - Private Ponds & Irrigation Wells – (High Demand)
  - Permitted Flowing – Surface Water – (Variable Demand or Availability)

3

POTW Effluent – Treated Waste (No Demand)

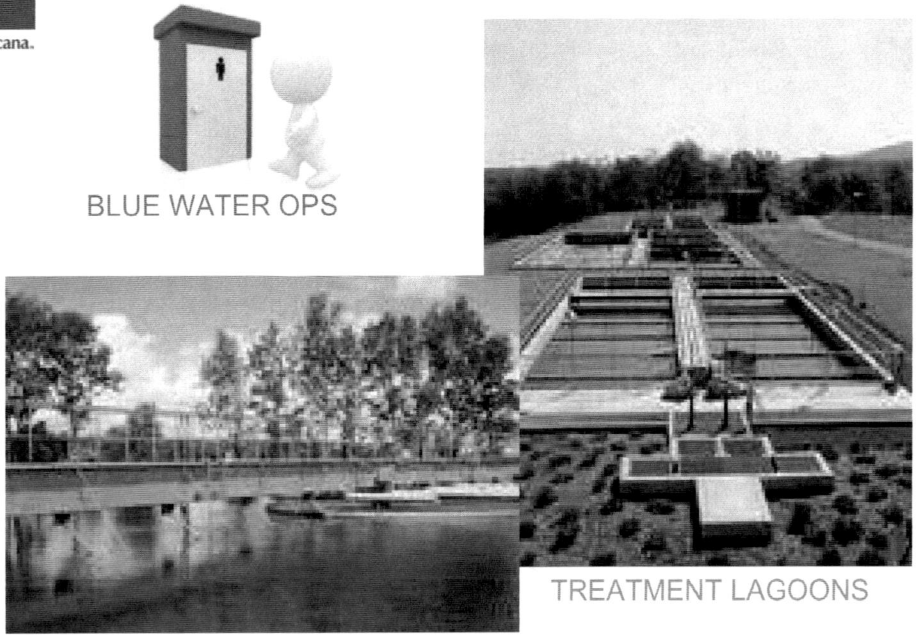

4

Brine / Brackish Water (No Significant Demand)

- USGS Map of Saline Water Formations

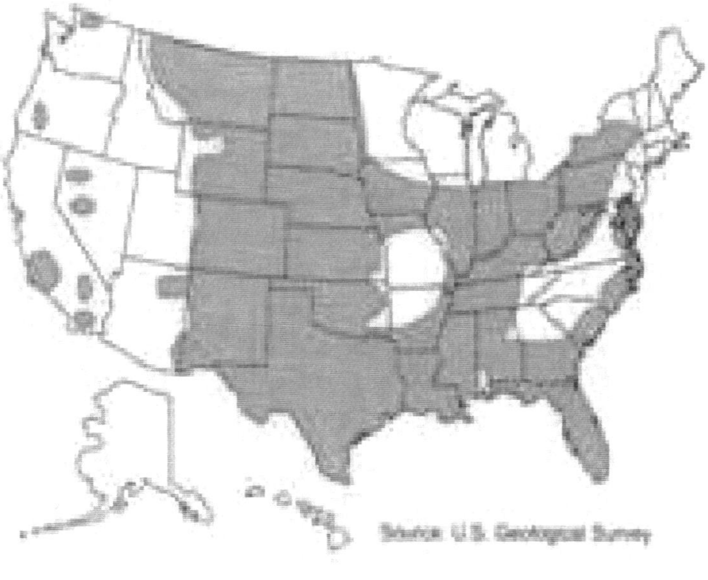

5

Produced Water / E&P Liquids – (No Demand)

encana.

SAVE THE SWD

6

Permitted Flowing / Surface Water – (Variable Demand or Availability)

encana.

7

encanа.

Hydraulic Fracturing
Relative Water Requirements per Well by Play

- MSL (Kansas) 46 - 50M
- MSL (Osage) 46 - 50M
- TMS 215 – 315M
- EBN 125 – 225M
- NLA 230 – 315M

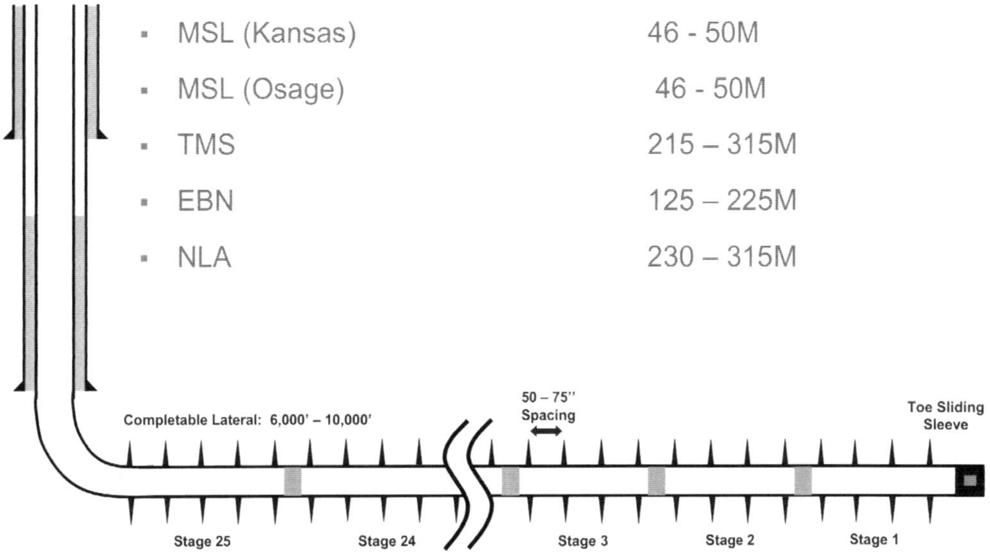

Typical Stage Count: 20 - 30

Note: Diagram not to scale

8

encana.

Frac Ops

9

Water Sourcing Options – Set Your Order of Preferred Use

1. Alternative Fresh Water Sources
   - POTW Effluent – Treated Waste (No Demand)
   - Alluvial Water Wells (No Significant Demand)
2. Non-Fresh Water Sources
   - Brine / Brackish Water (No Significant Demand)
3. Reused / Repurposed Water
   - Produced Water – (No Demand)
   - Flowback - (No Demand)
   - E&P Liquids from Drilling Operations - (No Demand)
4. Fresh Water Sources
   - Private Ponds & Irrigation Wells – (High Demand)
   - Permitted Flowing – Surface Water – (Variable Demand or Availability)

10

Advice for the Water Resources Schmuck – ALWAYS

- Have a Trained landman source your water
- Expect timely, exact answers and massive detail from your sourcer. He must be legally, logically & situationally correct. Only keep the good ones.
- FIRE the landman that does not meet your expectations. They won't improve
- PAY for your source water – nothing is free
- Pay quickly, follow up quickly
- Have multiple sourcing options (2 fresh, 2 alternate)
- Have multiple ways of getting the water from source to site
- Limit the number of frac tanks used to store water
- Limit the number of reps talking to the lease & surface owners
- Limit HAULING
- Communicate (face/phone/email/text) you cannot over communicate
- Beware of terminology differences between disciplines
- Check, double check, triple check (ask: what did you hear me say?)
- Use multiple transfer companies
- Use local talent as much as possible

11

encana.

Advice for the Water Resources Schmuck – NEVER

- Let a water transfer outfit source your water
- Let Drilling select your water source
- Let the sourcer give vague answers or put you off
- Let your transfer company talk to your surface owners
- Drill a water well, improve a pond or dig a pit for a surface owner
- Get too close with your transfer company
- Depend on a single government/NGO permitted water source
- Have only 1 water source
- Make it easy on Supply Management – If they like it, you are doing something wrong

12

encana.

THAT'S ALL FOLKS

Sam Bacon
MCBU Water Resources Lead
Encana Oil & Gas (USA) Inc
sam.bacon@encana.com

13

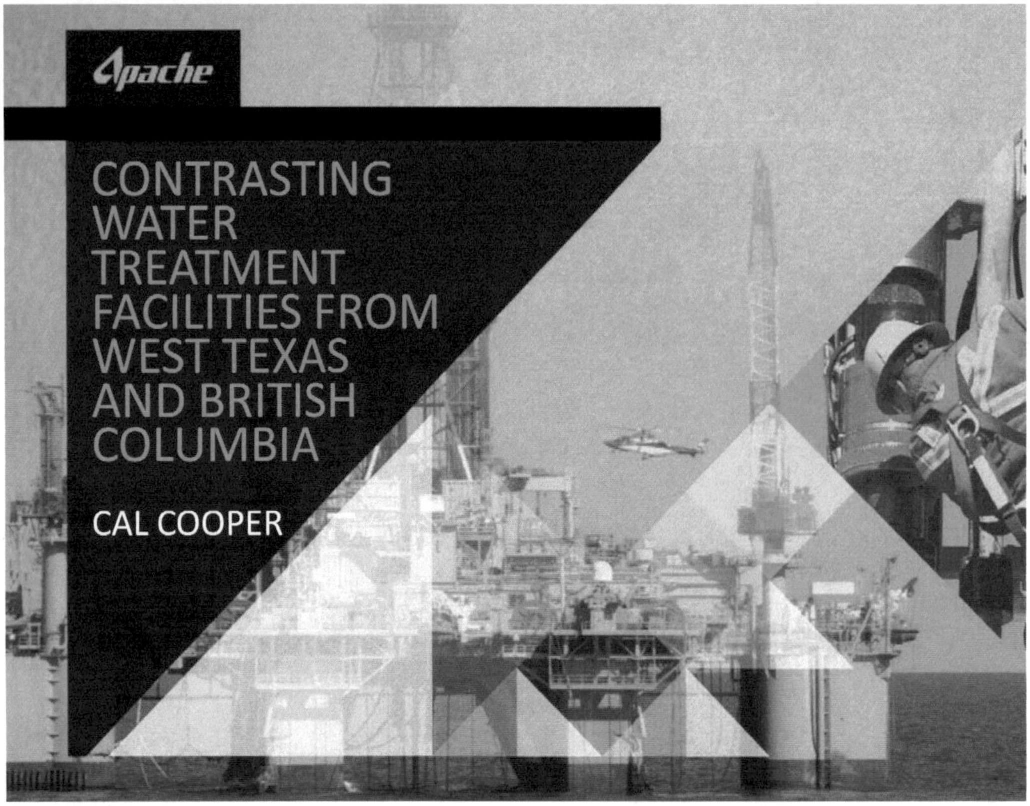

WATER DRIVERS NO WATER = NO FRAC

- Regulatory & Environmental constraints
- Availability - Options
- Operational Choices
- Water Quality Needs
- Treatment required
- Technology / Cost
- Bottom line

2 Apache

SURFACE WATER & SHALLOW AQUIFERS

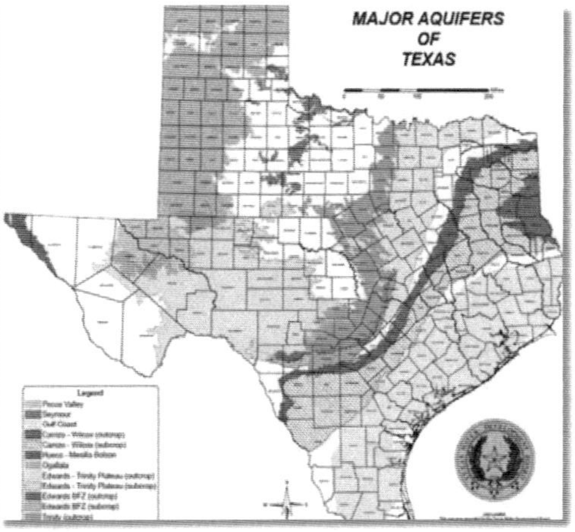

- Great when you can get it.

3

- Bug factories
- Simple treatments
- No HF performance issues

4

WATER QUALITY – OPERATIONAL REQUIREMENTS

- Fresh or saline?
- TDS (<35,000)
- Fe (< 10 ppm)
- Bug level (low)
- No Barium, CaCO3 sulfate issues
- pH (mid range)
- Timeframe

5

PRODUCED WATER PROBLEMS

- Safety
- Compatibility
- Infrastructure
- Disposal – $2.50+/bbl

6

PRODUCED WATER CHALLENGES

DO YOU KNOW WHAT'S IN YOUR WATER?
IS IT STABLE OR REACTING?

- Hydrocarbons
- Biologicals
- H_2S
- TDS
- pH
- Fe, Ba, Mg, Mn, CaCo3, Sulfates

7

Apache

PRODUCED WATER TREATMENT TECHNOLOGIES

1. Chemicals
2. Ozone oxidation
3. Nano-filtration
4. Hydrocyclones
5. Deionization
6. UV

MVR Evaporator, RO, EC…and many more

8

Apache

CHLORINE DIOXIDE

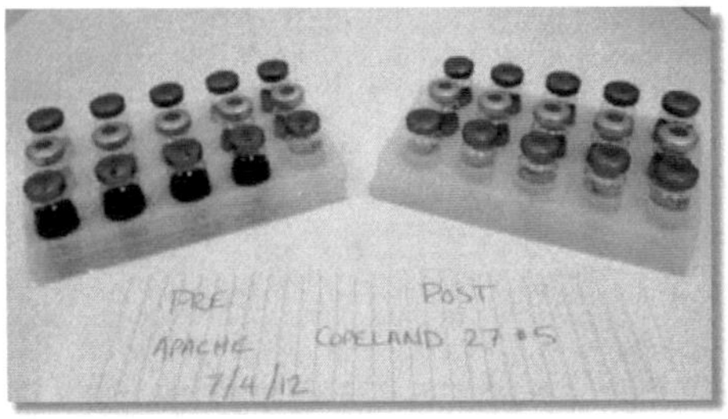

Apache 1st produced-water frac

- Deadwood area - Now > 12
- 27,000 bbls treated
- Iron reduction to 6 ppm
- 0 bug bottles

9

CHLORINE DIOXIDE

- Side View:

10

CHLORINE DIOXIDE

Advantages

- Kills bugs
- Scavenges sulfides (FeS, H_2S)
- No colloidal sulfur
- Very small footprint
- No toxic byproducts
- Shown to separate oil & water

Limitations

- Limited residual effect
- pH 6-10
- Increased chlorides
- Work-intensive

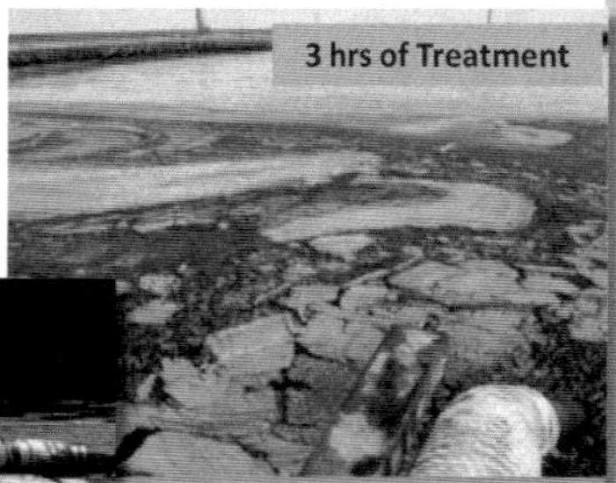

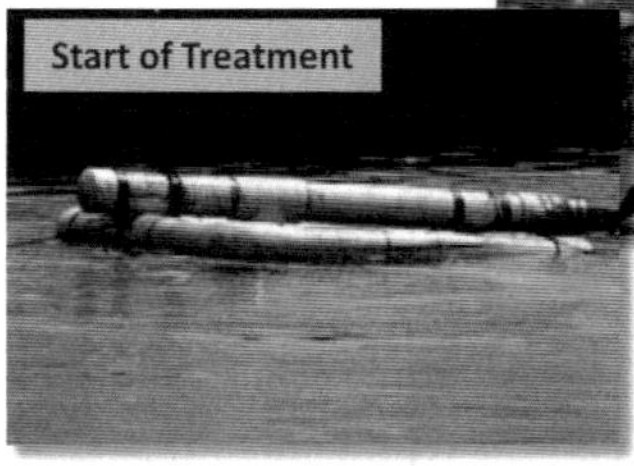

11

Apache

pH TREATMENTS

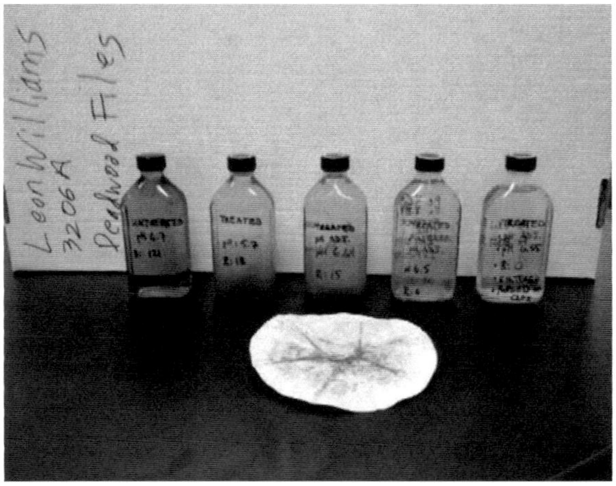

- Treat ClO_2
- Raise pH w Caustic
- Pptn Fe++
- Drops pH
- Filter
- Purge ClO_2

12

Apache

ORGANOPHOBIC NANO-TECHNOLOGY FILTRATION

MOLECULAR FILTRATION & LANCE ENERGY

Advantages

- Separates bugs, organics, & Fe
- No biofouling = long-lasting filters
- Low pressure – max 4 bar
- No heat produced
- No chemical additives
- Clear effluent
- Easy to use

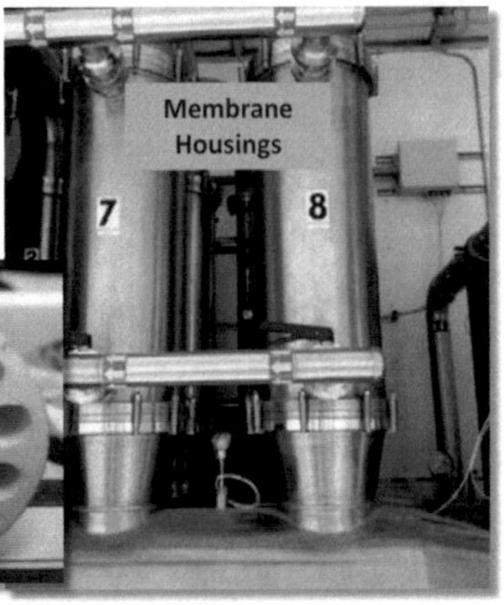

Limitations

- 2,000 bpd
- No continued treatment
- No ion reduction
- R & D stage

Apache

13

APACHE TEST – FLOWBACK WATER TREATMENT

ORGANOPHOBIC NANO-TECHNOLOGY FILTRATION

Lance Energy Bug Test Results

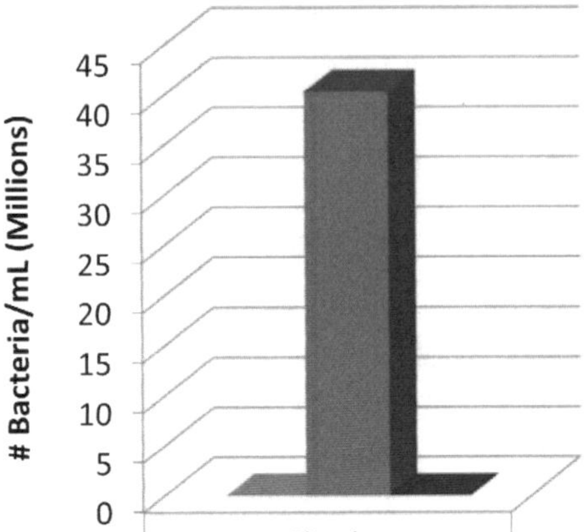

	Clinch
Fresh Water	20,381
Pre-Treat	40,451,059
Post-Treat	12,872

- Apache BSE test
  - Significant bug removal
  - Clear permeate
- Continued development

Apache

14

SALINE AQUIFERS

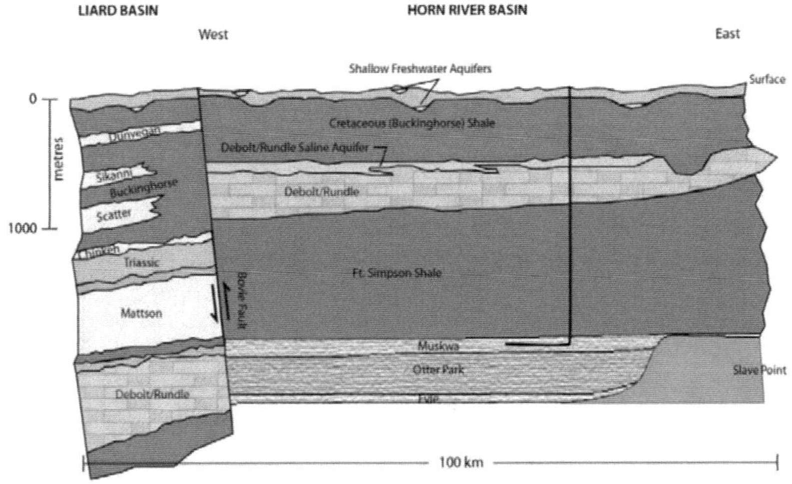

Figure 1. Schematic stratigraphic cross-section, Horn River Basin and adjacent Liard Basin, modified from Petrel Robertson.

- Quality
- Flow Rate
- Depth
- Who owns it?
- Limited data
- Surface Facilities

15

DEBOLT WATER TREATMENT

16

Apache

WATER AND COMPRESSION PLANTS

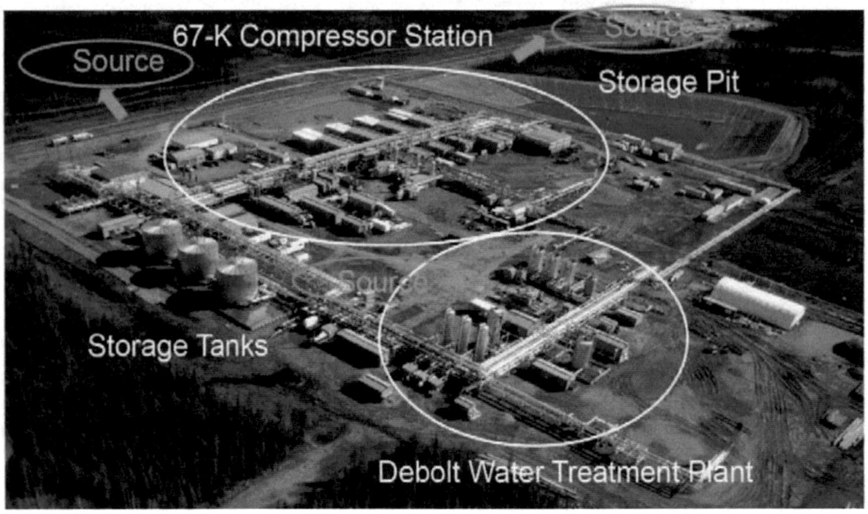

17

DEBOLT FRAC WATER SYSTEM

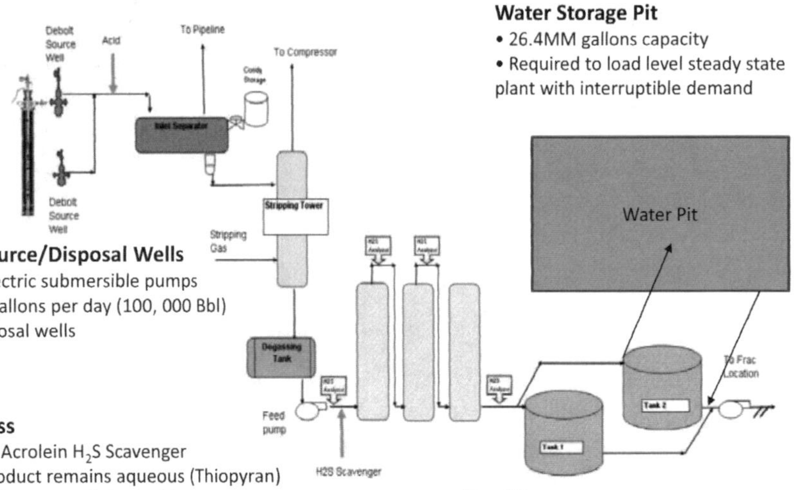

Water Storage Pit

- 26.4MM gallons capacity
- Required to load level steady state plant with interruptible demand

Water Source/Disposal Wells

- Three electric submersible pumps
- 4.2MM gallons per day (100, 000 Bbl)
- Two disposal wells

Process

- Uses Acrolein H_2S Scavenger
- Byproduct remains aqueous (Thiopyran)
- Inlet Separator, Stripping Tower and Retention Vessels make up the process
- 3 H_2S Analyzers downstream of chemical injection measure in ppb in the water phase

Frac Water

- Water is free of H_2S
- Temperature is >100F, excellent for winter operations
- 4 FVD driven centrifugal pumps provide +100BBL/min to site

Courtesy Brad Affleck

18

STRIPPING AND REACTION TOWERS

19

Apache

THANKS!

20

Apache

OASYS WATER

The Oasys® Forward Osmosis Membrane Brine Concentrator:

Commercialization of FO technology for brine minimization

Nathan T. Hancock, Ph.D.
Director, R&D

AIChE & AWMA Joint Workshop: Shale Oil and Gas E&P
November 1-2, 2012
Pittsburgh, PA

Presentation Overview

- Forward Osmosis (FO)
  - Concept
  - Oasys Membrane Brine Concentrator
- Pilot-scale studies
  - Marcellus shale
  - Permian basin
- Conclusions

RO: Fighting Osmotic Pressure

$$J_w = A(\Delta P - \Delta\pi)$$

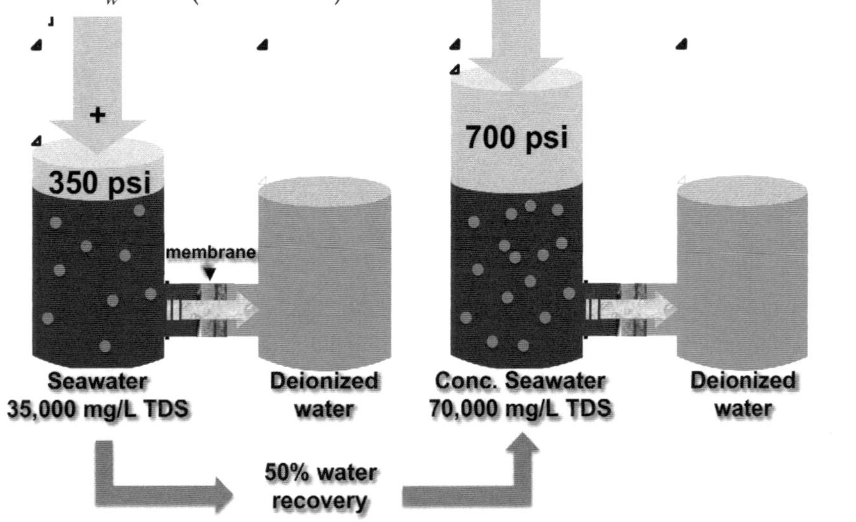

WWW.OASYSWATER.COM 2 Private and Confidential

Engineering Osmosis

$$J_w = A(\Delta P - \Delta\pi)$$

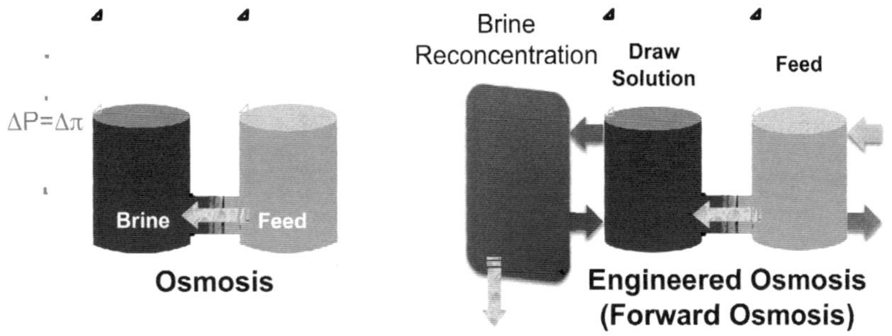

WWW.OASYSWATER.COM 3 Private and Confidential

Materials: Oasys Proprietary Draw Solution

Highly soluble thermolytic draw solutions
- Solubility exceeding 30% mol/mol H_2O
  (Compare to NaCl at ~10% mol/mol H_2O)

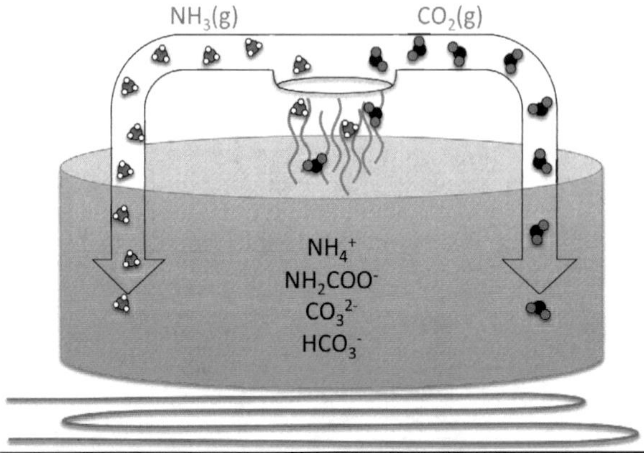

WWW.OASYSWATER.COM 4 Private and Confidential

Introducing the Oasys Membrane Brine Concentrator

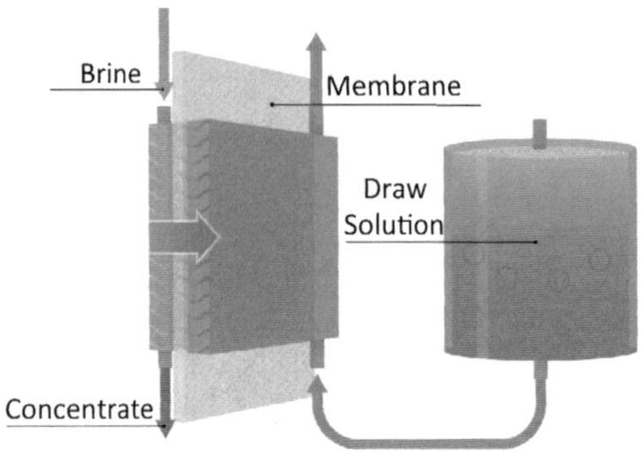

WWW.OASYSWATER.COM 5 Private and Confidential

Introducing the Oasys Membrane Brine Concentrator

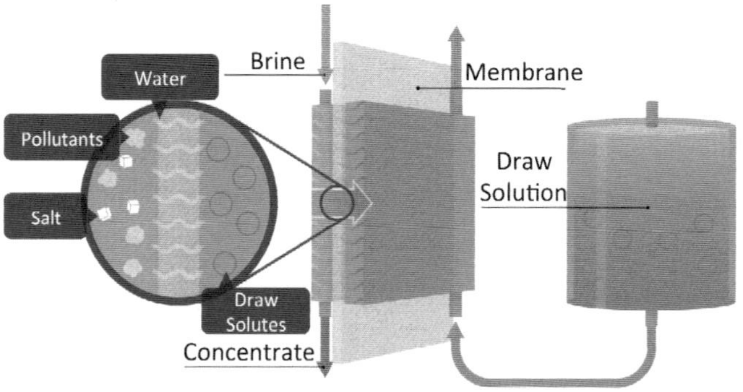

Introducing the Oasys Membrane Brine Concentrator

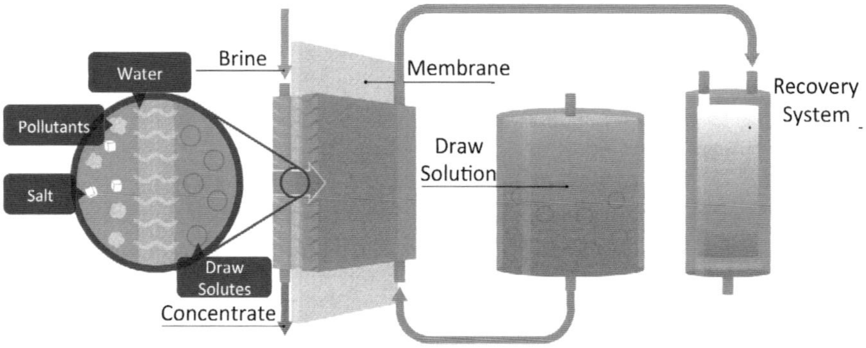

Introducing the Oasys Membrane Brine Concentrator

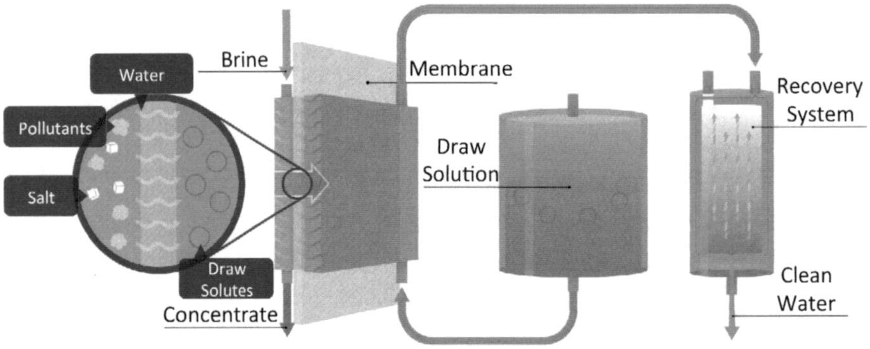

WWW.OASYSWATER.COM 8 Private and Confidential

Introducing the Oasys Membrane Brine Concentrator

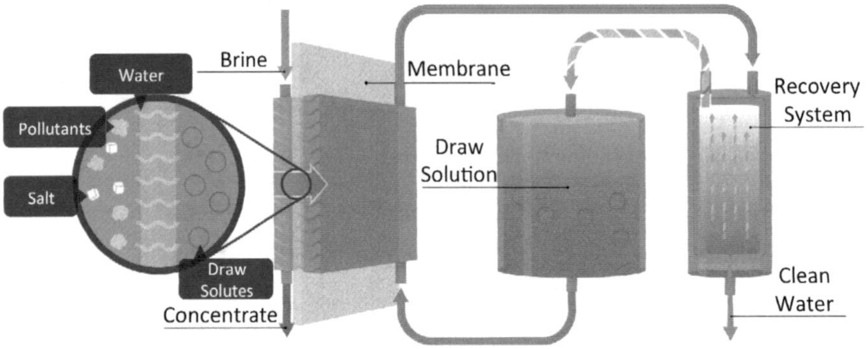

WWW.OASYSWATER.COM 9 Private and Confidential

Introducing the Oasys Membrane Brine Concentrator

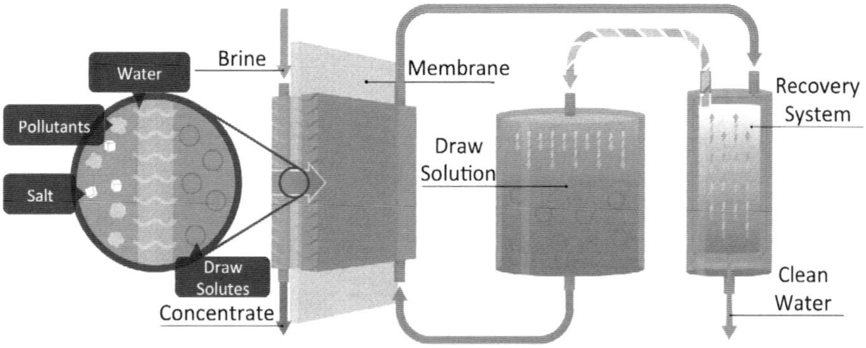

Materials: Oasys FO Membrane

Industry leading TFC FO membrane module

Parameter	Value	Unit
Pure water permeability (A)	$1.02 \cdot 10^{-11}$	m/s-Pa
NaCl permeability (B)	$5.74 \cdot 10^{-8}$	m/s
Structural parameter (S)	356	µm
FO mode water flux (J_w @ 1 M NaCl vs. DI)	16.4	GFD
NaCl rejection (RO mode; 225 psig; 2,000 ppm)	99.1%	

Oasys FO MBC System Piloting: Marcellus Produced Water

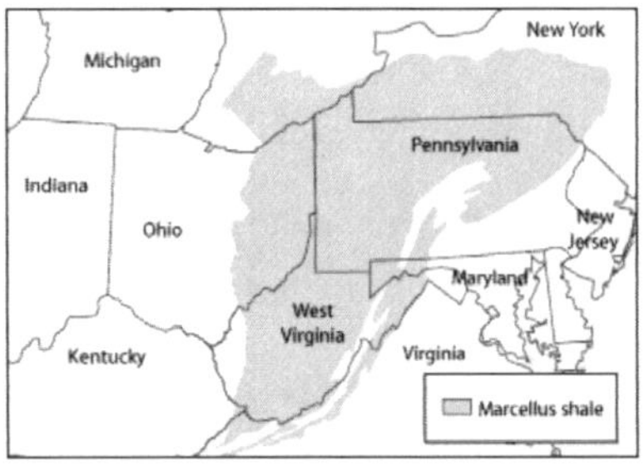

Image source: www.NYC.gov

Pilot Unit Overview

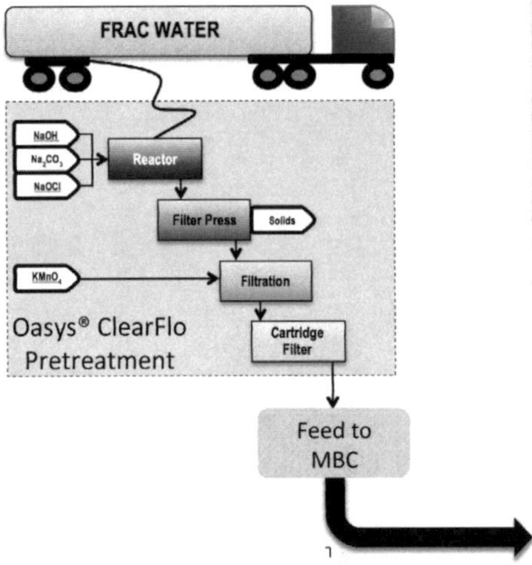

Synopsis of Study – Marcellus

- Pilot Oasys FO MBC operated for 800+ hours
    - 24-hour continuous and daily startup and shutdown modes demonstrated
- Treated 60,000+ gallons of frac flowback and formation water from the Marcellus shale gas play

Raw Feed Water Characteristics

TDS: ~70,000 mg/L
Hardness: ~15,000 mg/L – $CaCO_3$
Strontium: 1,300 mg/L
Silica: 30 mg/L

TOC: ~120 mg/L
Gross Alpha: 170 pCi/L
Bromide: 250 mg/L
Boron: 10 mg/L

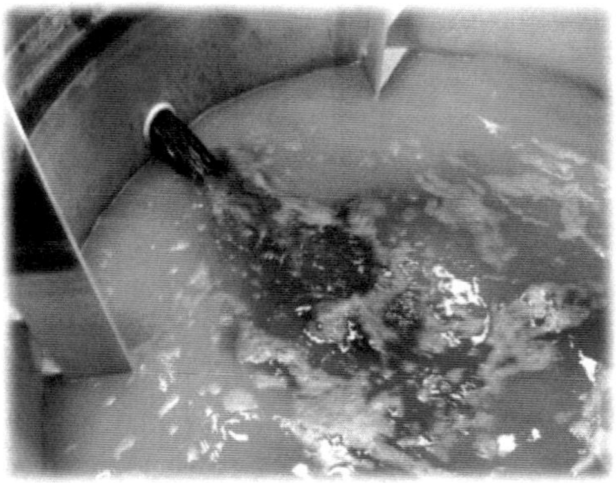

Summary of Feed to Oasys MBC

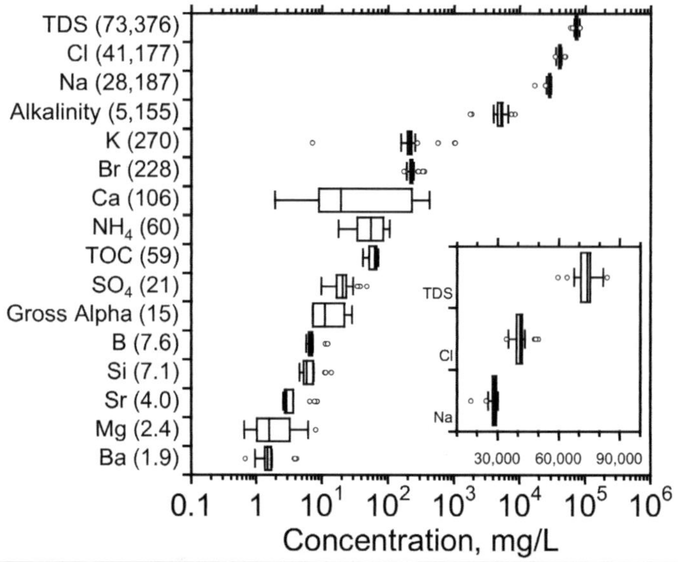

Process Data: Water Flux and Recovery

- Sustained 65% water recovery of 73,000 mg/L TDS feed with FO

- No makeup draw solutes added
- Minor fluctuations due to non-optimized boiler cycling

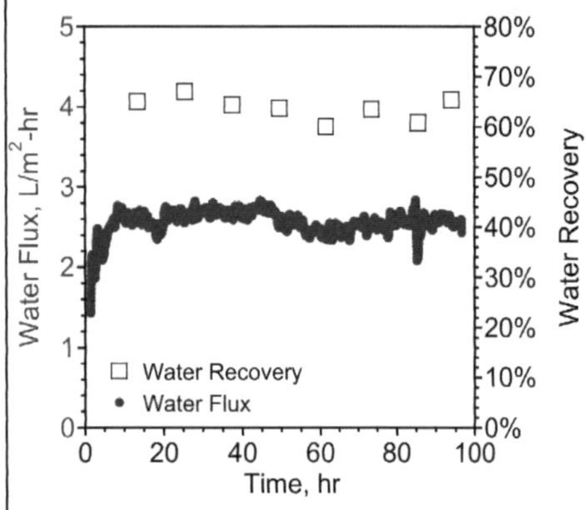

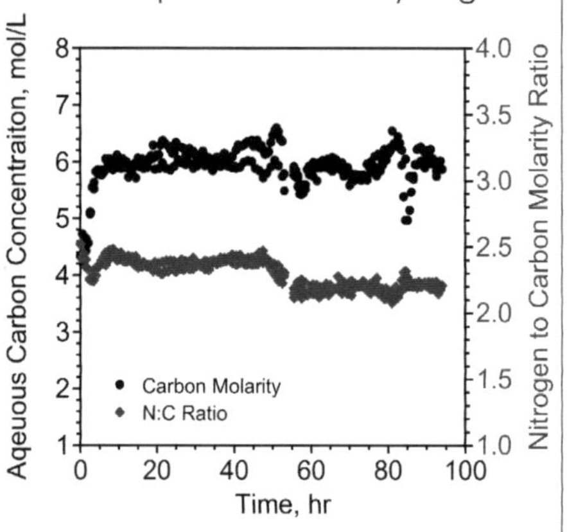

Cumulative Pilot Data: Concentrate

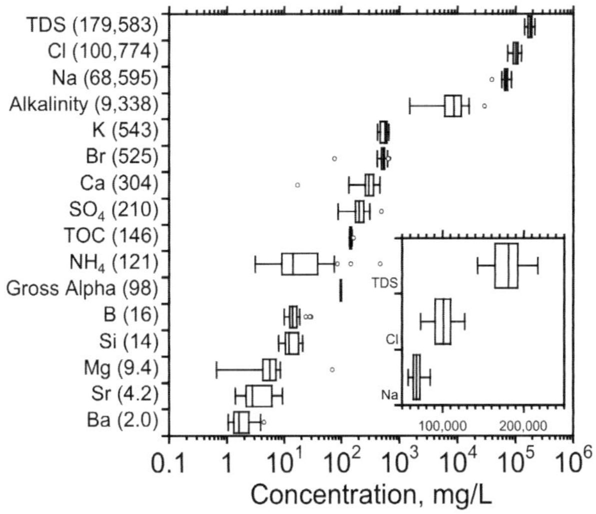

Cumulative Pilot Data: Product Water

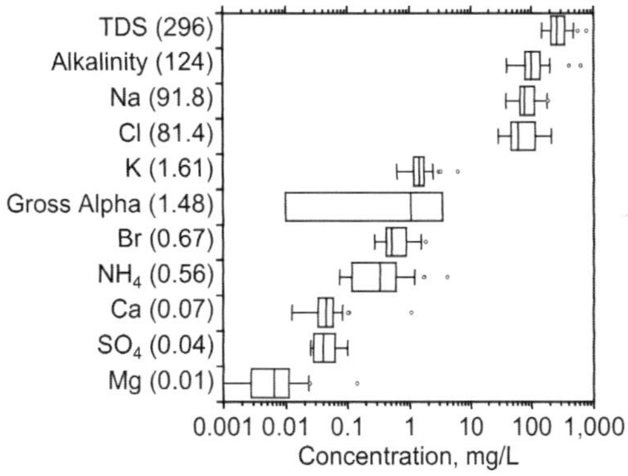

Boron removed to <0.05 ppm

Oasys FO MBC
System Piloting: Permian (*ongoing*)

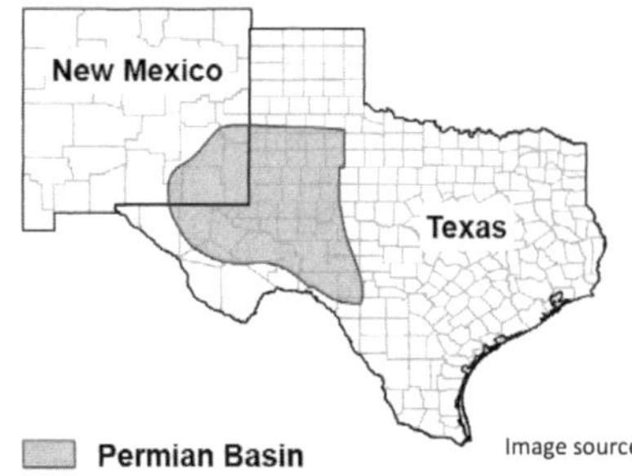

SELECT ENERGY SERVICES

Image source: www.circlestarenergy.com

WWW.OASYSWATER.COM 20 Private and Confidential

Permian Study to Date

- Pilot Oasys FO MBC operating for 350+ hours
  - Daily startup and shutdown
  - 24-hour continuous
- Treating 37,000+ gallons
  - Frac-flowback and produced water from the Permian basin

WWW.OASYSWATER.COM 21 Private and Confidential

Permian Pilot Data

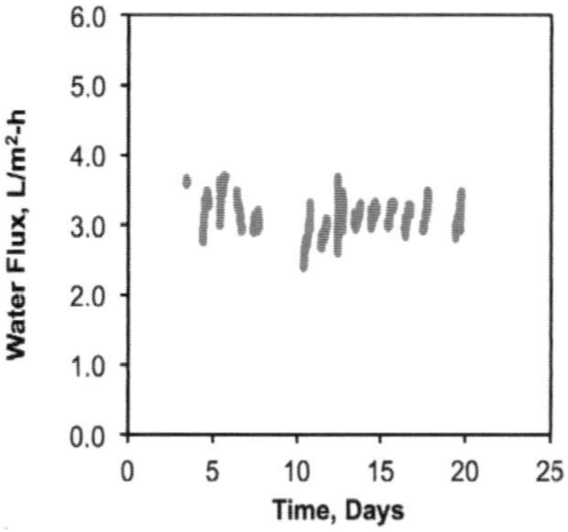

Average Water Recovery: 60%

10 lb brine,
valuable to producers

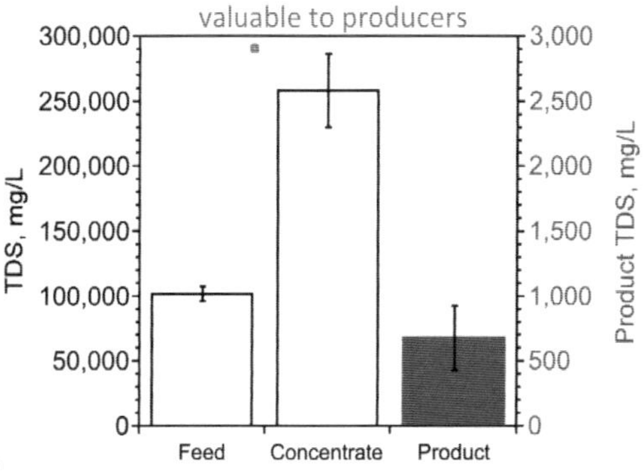

Open Cycle Energy Consumption

Oasys® MBC Pilot

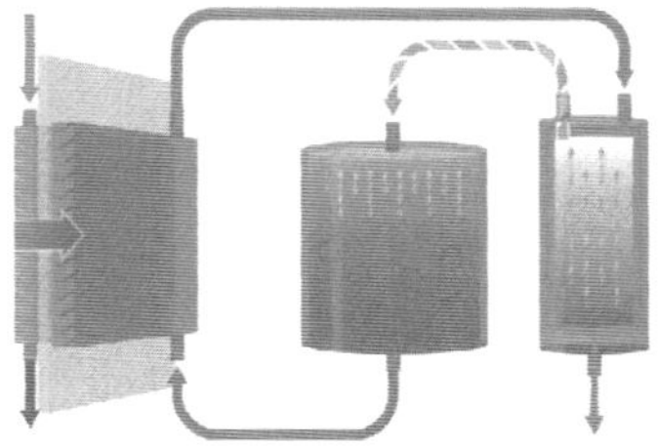

VS.

Single Stage Evaporator

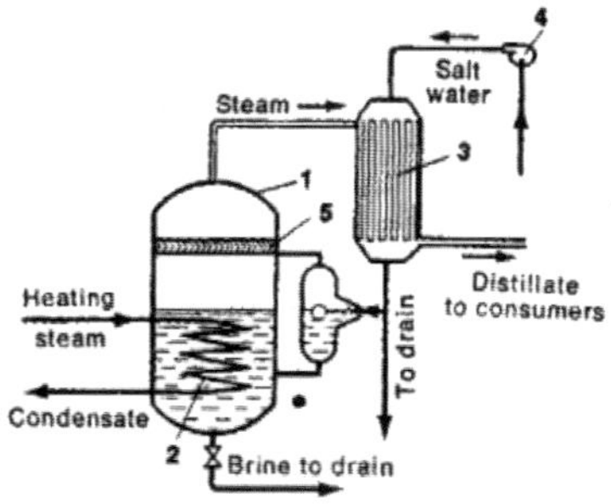

~275 $Wh_{thermal}$/kg product
8.5 $Wh_{electric}$/kg product

Oasys MBC
~55% Less

~630 $Wh_{thermal}$/kg product
1–8 $Wh_{electric}$/kg product

… But nobody employs commercial evaporative distillation systems this way

Oasys MBC 4000

System Layout

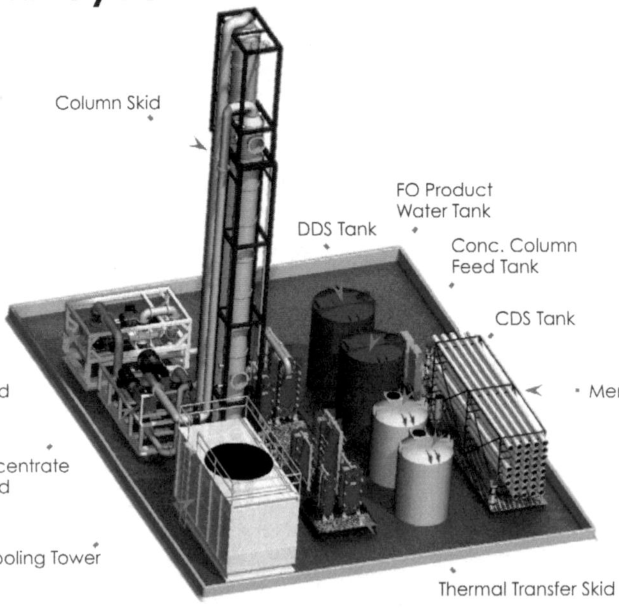

Mechanical Vapor Compression

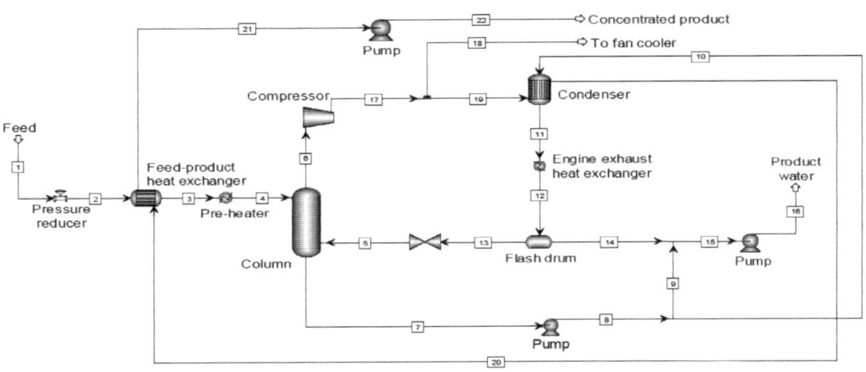

Oasys MBC 4000 energy consumption between 3.5 and 5 kWh/bbl treated, depending on feed salinity and recovery

WWW.OASYSWATER.COM Private and Confidential

Conclusions

- Oasys FO MBC
  - High recovery, low TDS product water
  - Reduces volume of brine for disposal, or…
    - Concentrated (heavy) brine has a variety of uses in O&G
    - Waste stream transformed to revenue source
  - Separation at ambient temperature and pressure reduces capital component cost
  - Vaporization of solute rather than solvent enables energy reduction for treatment
- Although FO systems are more robust, good pretreatment is still important
- Design and demonstration of an Oasys FO system for renewable desalination underway

WWW.OASYSWATER.COM 27 Private and Confidential

OASYS
WATER

Nathan T. Hancock, Ph.D.
Director, R&D
nhancock@oasyswater.com
www.linkedin.com/in/NTHancock
@NTHancock

www.OasysWater.com
@oasyswater

WWW.OASYSWATER.COM 28 Private and Confidential

Treating and Reusing Flowback and Produced Water

HECKMANN
CORPORATION

Discussion Topics

- Volume and Characteristics of Drilling, Flowback, and Produced Water
- Flowback and Produced Water Treatment and Reuse

HECKMANN
CORPORATION

Volumes and Characteristics of Oil/Gas wastewaters

- Top Hole drilling water, rainwater, etc- low volumes (~2000bbl/well), low TSS and low TDS
- Intermediate and bottom hole drilling water-low volumes, high TSS and low TDS
- Flowback water volume typically 10-40% of frac water (~10 x drilling water volume)-low TSS, high TDS
- Produced water-lower volumes than flowback, low TSS, high TDS

HECKMANN CORPORATION

3

Typical Makeup of Flowback and Produced Waters

- Mainly Water
- Total dissolved solids (TDS)
- A mixture of carbonates, chlorides, sulfates, nitrates, sodium and other materials
- Oils and greases
- Low levels of NORM depending on formation

HECKMANN CORPORATION

USGS Inlet TDS Survey

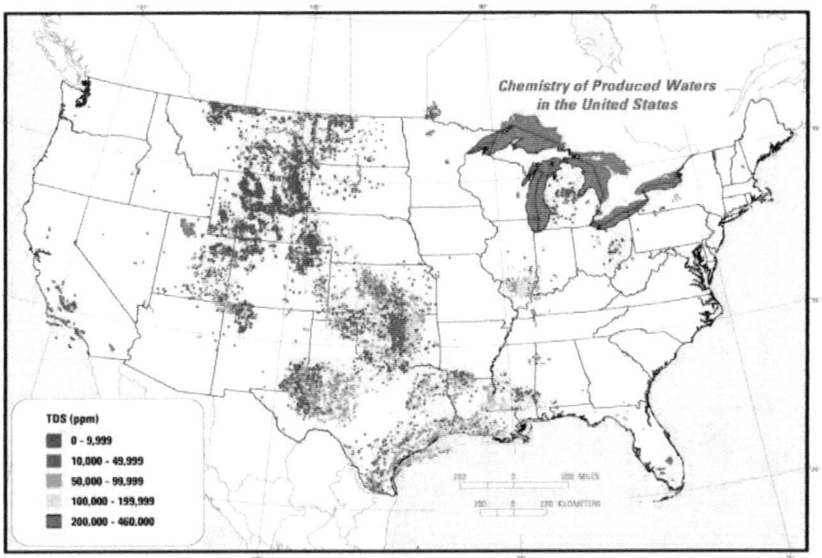

Source: http://water.usgs.gov/orh/nrwww/Otten.pdf

HECKMANN
CORPORATION

What will the producers accept as recycled products?

- Filtered/disinfected low TDS waters as "fresh water" for hydrofracturing
- Treated Flowback and Produced water with oil, TSS, scalants removed, and disinfected for blending with fresh water
- Hydrocarbons recovered from hydrocarbon based drilling muds
- Cleaned cuttings and well pad gravel for reuse
- Calcium chloride, barite, heavy brines that can be produced from the flowback and produced water
- Distilled water from thermal treatment systems

6

HECKMANN
CORPORATION

Reuse logistics

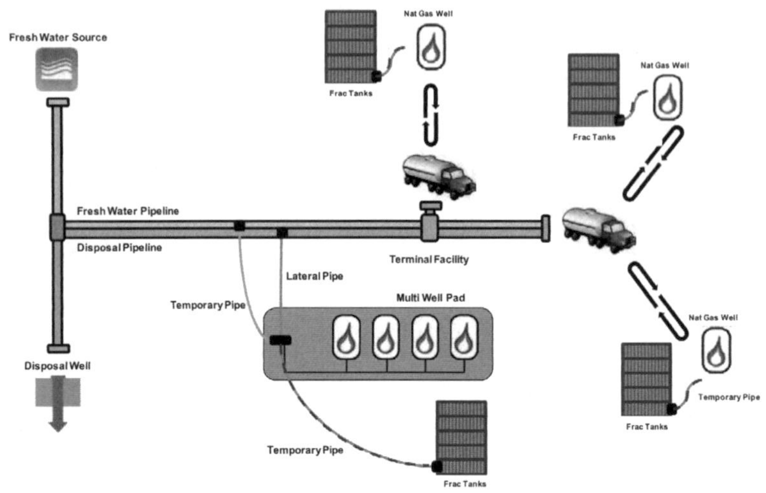

HECKMANN
CORPORATION

7

Challenges in treatment selection

- Economics
- TDS, TSS variations from field to field
- Frac chemical variation
- Oil and other organics present
- Intended use for treated water – reuse or disposal
- Mobile vs. fixed assets
- Producer owns and operates vs. working with service provider

HECKMANN
CORPORATION

Typical low TSS, low TDS treatment

- Onsite or central treatment

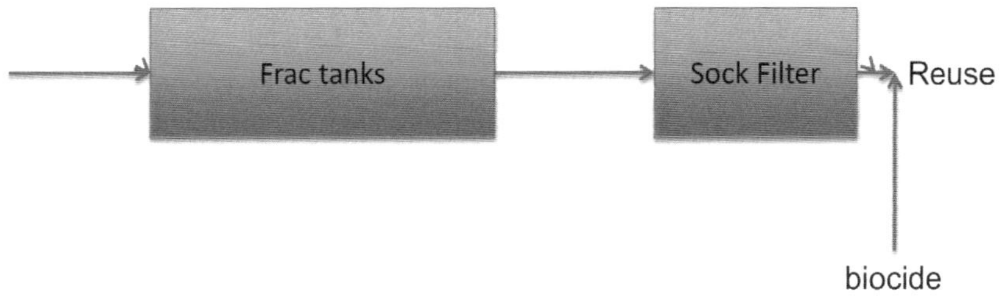

HECKMANN CORPORATION

Typical high TSS, low TDS treatment

- Onsite or central treatment

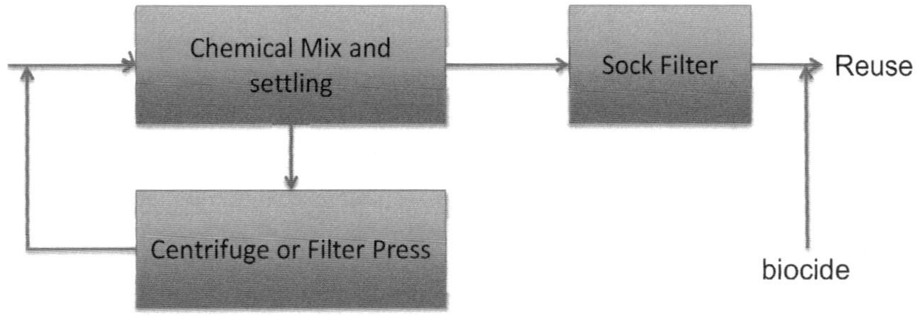

HECKMANN CORPORATION

Flowback/Produced Water "Floc and Drop" Treatment

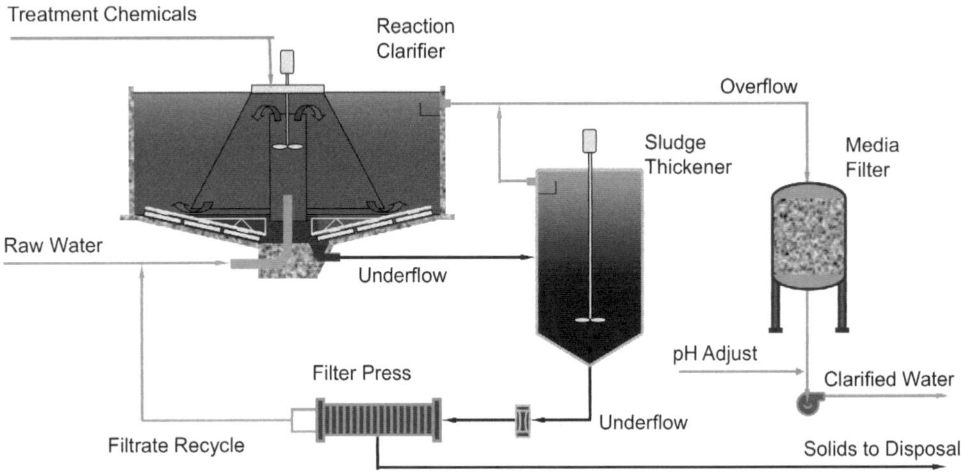

HECKMANN CORPORATION

Source: VWS Oil and Gas

Flowback/Produced Water Electrocoagulation System

12

HECKMANN CORPORATION

Flowback/Produced Water Treatment-EC system

- Halliburton EC system results

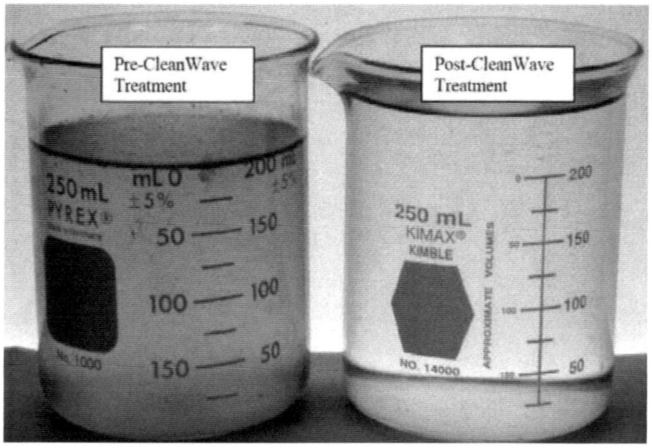

HECKMANN CORPORATION

13

Need to remove oil in the liquids fields

- Free Oil Removal Process
- Froth Floatation Technology
- Mechanical Gas Induction
- 90 ~ 95% Free Oil Removal
- Effective De-Oiling Process

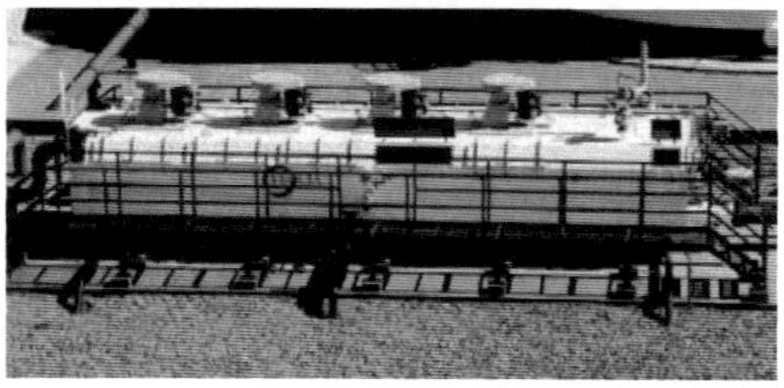

Source: VWS Oil and Gas

HECKMANN CORPORATION

Total Dissolved Solids (TDS) Removal/Concentration

Limited Application in High TSS waters due to osmotic pressure limitations, pretreatment,cost and need to dispose or reuse concentrate

- Reverse Osmosis
- Forward Osmosis
- Membrane distillation

HECKMANN CORPORATION

Brine Concentrators

- High TDS Applications
  - > 3% Total Dissolved Solids
- Evaporators
  - Natural circulation
  - Rising film
  - Falling film
  - Forced circulation evaporators
- Crystallizers
  - Enhanced Forced Circulation
  - Forced circulation crystallization
  - Falling film crystallizers

Source: GE Water

HECKMANN CORPORATION

The ultimate recycle/reuse system-can we design it to be cost effective?

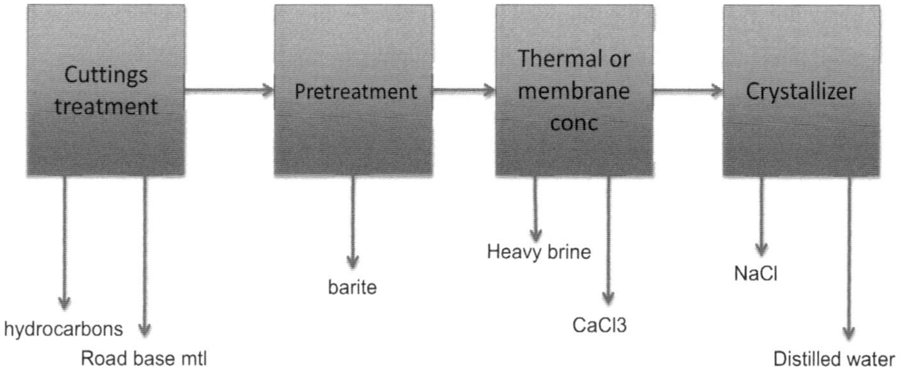

17

HECKMANN CORPORATION

Sorbents to Extract Value from Flow Back Water and Produced Water: Reversible Capture of Organics and Rare Earth Elements

Paul L. Edmiston
College of Wooster

Contact Information:
pedmiston@wooster.edu

AIChE – A&WMA Joint Workshop
Shale Oil and Gas E&P – Water Challenges and Opportunities

Sol-Gel Derived Mesoporous Silicas

Sol-Gel Process

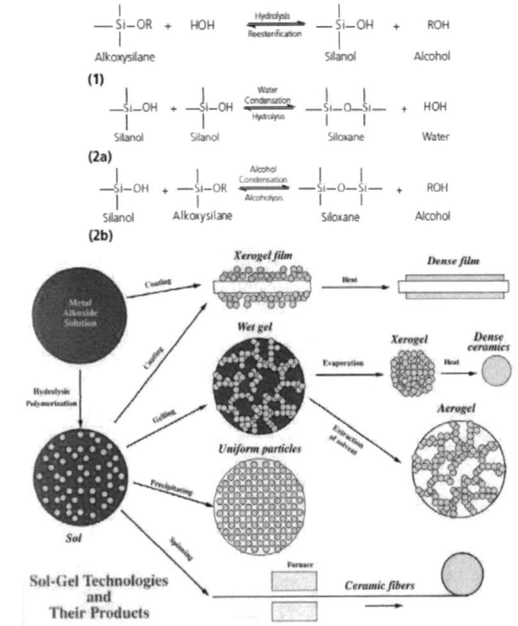

Ordered Templated Materials

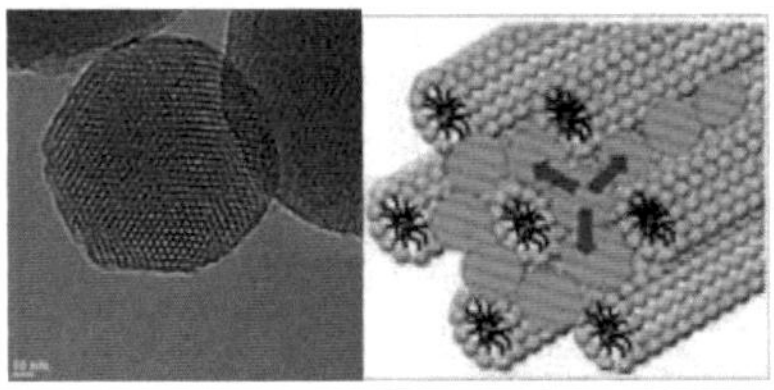

Aerogels

Polysilsesquioxanes

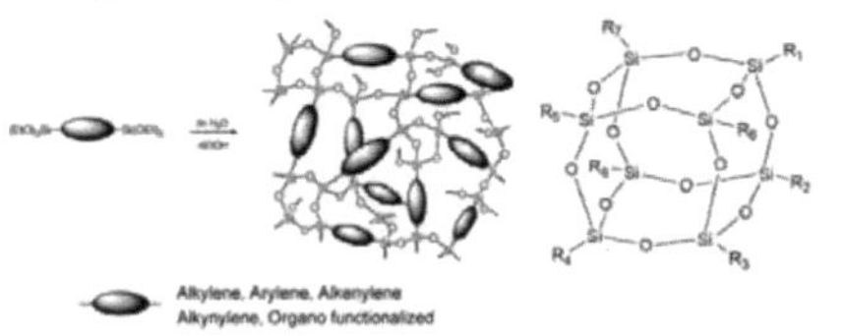

What is Osorb?

Osorb is a highly structured glass which instantaneously swells when it comes in contact with a wide range of organic molecules.

Absorbs: Gasoline, Ethanol, Propane, Solvents, Pharmaceuticals

Osorb swelling upon drop-wise addition of acetone

Swelling is due nanomechanical expansion of a mesoporous silica matrix

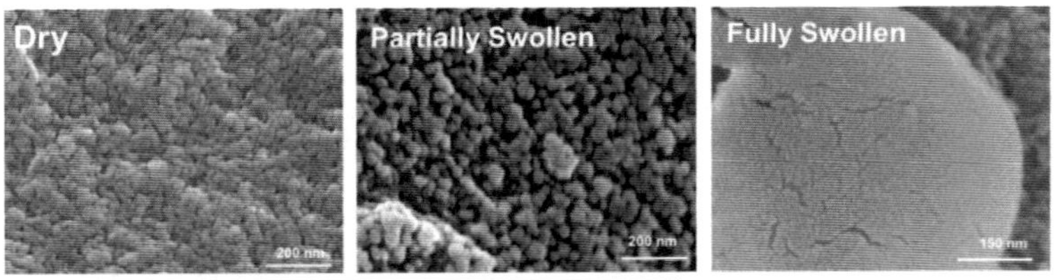

Osorb does not swell in water

Scanning Electron micrographs of Osorb swelling

Characteristics of Osorb

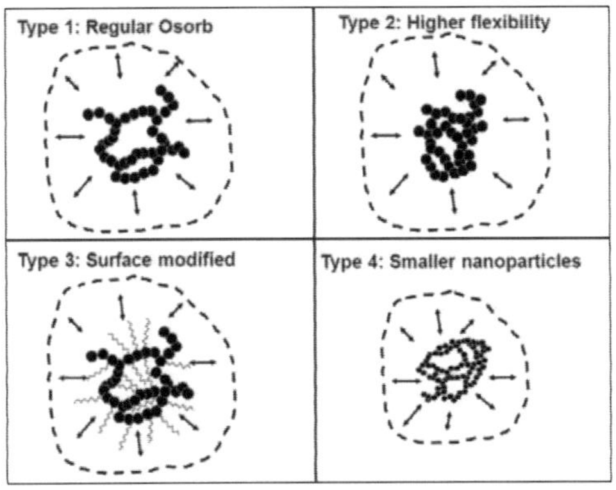

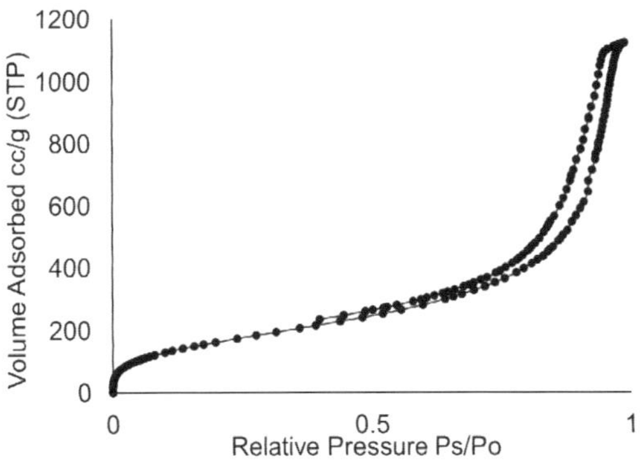

Surface Area and Pore Volumes of Various Osorb® Materials

Type	Swell mL/g	Surface Area(m^2/g)	Pore Volume (mL/g)	Pore Size Distribution (%) under 6 nm	6-8 nm	20-80 nm
1	5.2	885	2.85	6	8	68
2	9.8	416	0.57	48	22	-
3	4.6	171	0.27	98	-	-
4	2.5	803	0.98	20	15	38

Force Generation Upon Swelling

Organic liquids

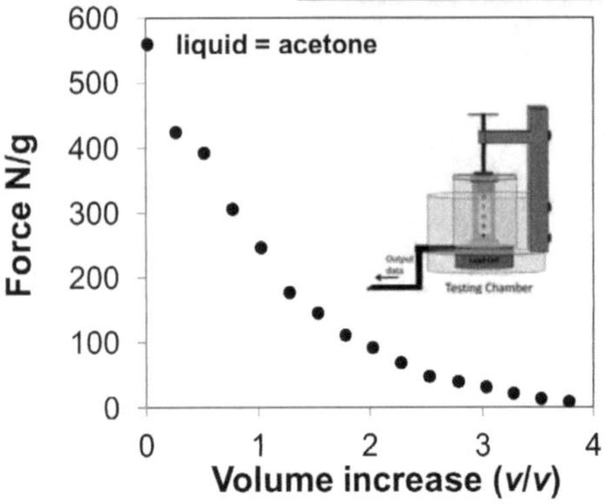

Max force 600 N/g (61,000 *w/w*)
Work = 0.8 ± 0.1 J/g
ΔH_{swell} = 5.2 ± 1.2 J/g

Entropically driven process
300% ΔV, 650% $\Delta mass$

Hydrocarbon vapors

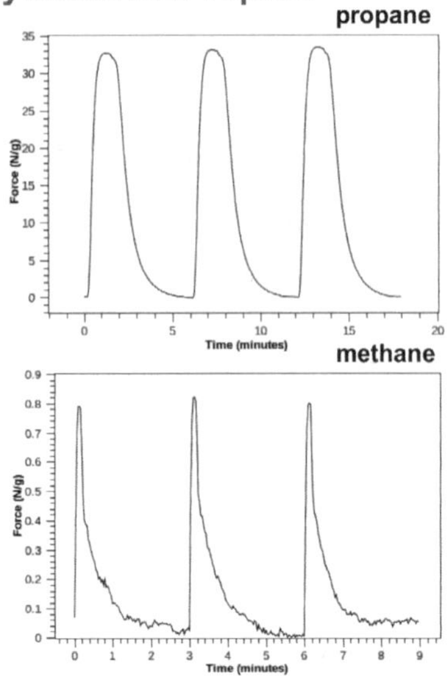

Max 1x *w/w* change for condensable vapors when $p=p_0$ 13% volume,

Produced Water Treatment

Osorb® removes a wide range of organics from water:

Compound	% Extraction
Aliphatics	
Straight chain	>99%
Branched chain	>99%
Cyclic	>99%
Aromatics	
BTEX	>98%
Polyaromatic	>99%[#]
Acids	
Carboxylic acids	40-95%, Depends on chain length
Napthenic acids	>90%
Aromatic acids	10-90% (depends on pH)
Phenol	>50% (pH =7.0) >90% pH
Surfactants	
Polyethylene oxide	>90%
Triton X-100	>90%
Alkyl glucosides	50-60%%
Other	
Metals	>80% extraction Fe^{2+}, Hg^{2+}, Ba^{2+} and Pb^{2+}
Polymers	up to 90% extraction depending in class (inquire)
Alcohols	4-90%, depends highly on salt concentration

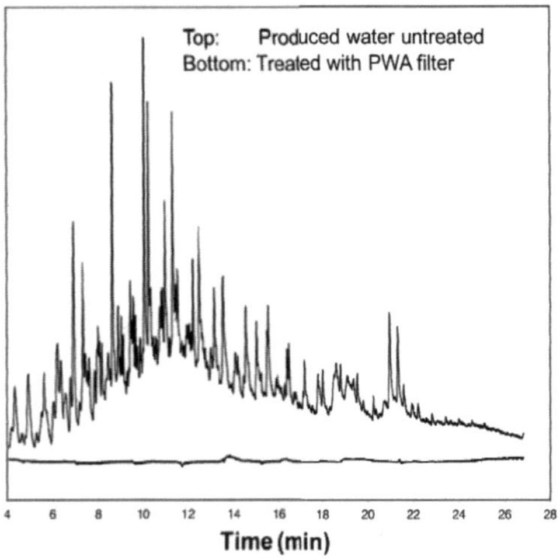

Absorption Model

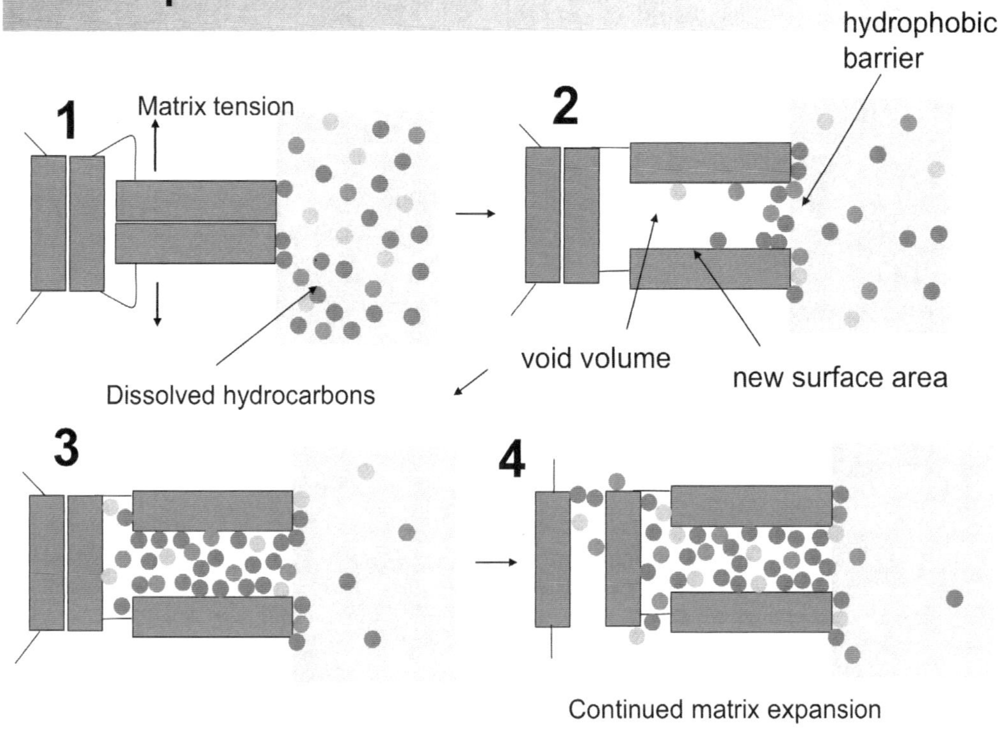

Absorption Model: Solid Solvent

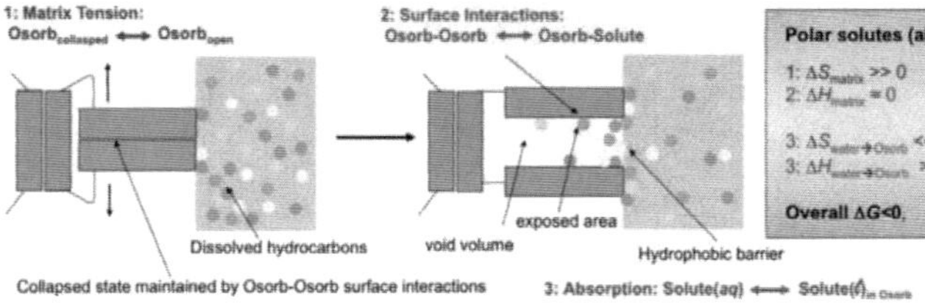

Polar solutes (alcohol, acids)	
1: $\Delta S_{matrix} >> 0$	(very favorable)
2: $\Delta H_{matrix} \approx 0$	(neutral)
3: $\Delta S_{water \to Osorb} << 0$	(very unfavorable)
3: $\Delta H_{water \to Osorb} > 0$	(unfavorable)
Overall $\Delta G<0$,	some absorption

Non-polar solutes (BTEX, TPH)	
1: $\Delta S_{matrix} >> 0$	(very favorable)
2: $\Delta H_{matrix} \approx 0$	(neutral)
3: $\Delta S_{water \to Osorb} < 0$	(unfavorable)
3: $\Delta H_{water \to Osorb} \approx 0$	(neutral)
Overall $\Delta G<<0$,	high absorption

Osorb® acts as a "solid solvent" that uses mechanical relaxation as an additional driving force for absorption of organics from water.

Expansion is endothermic indicating a decrease in entropy (ΔS_{matrix}) that is a significant energy term manifested by fact that swelling can produce mechanical forces that exceed 400N/g.

In general, there is a one order of magnitude greater partition coefficient for absorption by Osorb compared to liquid-liquid extraction due to the contribution from matrix expansion.

Extraction of 30 Compounds by Osorb vs. logK_{ow}

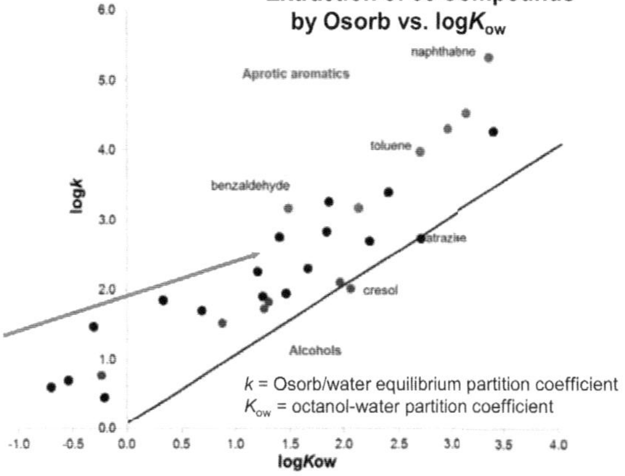

Conditions: contaminant concentration 100 ppm, 0.5% *w/v* Osorb per volume of solution, *T*=25°C.

Water Treatment Effectiveness Chart

Treatment	Applications
90-99%	Chemical EOR emulsions Dissolved BTEX and hydrocarbons Flow back water to remove biocides, slicking agents Slop water treatment to remove oil and non-ionic surfactants
50-90%	Solubles including cresols, phenols, C_3-C_{x+3} organic acids SAGD water, North Sea Alcohols from high TDS solutions
5-50%	Dissolved methanol Guar removal from flowback water Ionic surfacant removal from clean up operations
0%	Salt removal from any water stream

Treatment of Highly Impacted Water

Pesticide waste
Complex mixture of pesticides, dyes, BTEX, surfactants
(5% organics by weight)

Osorb-EB: Results on Tough to Treat Water

Emulsified Oil

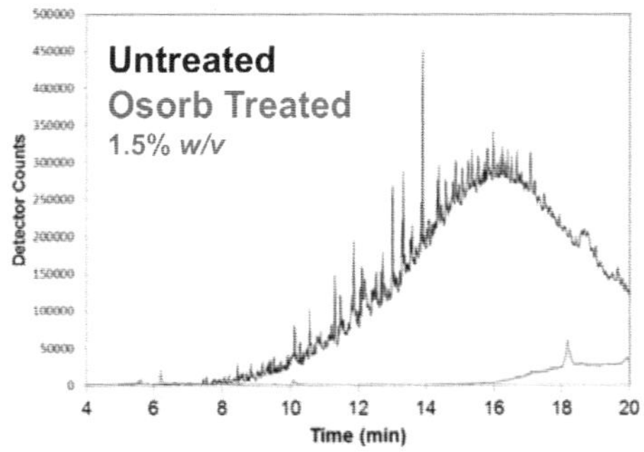

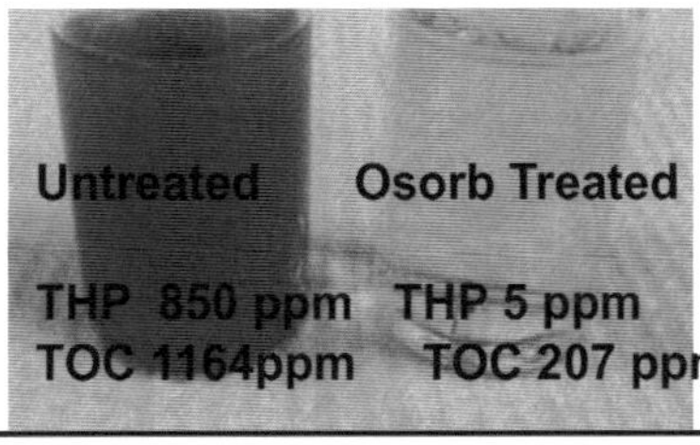

Flow back

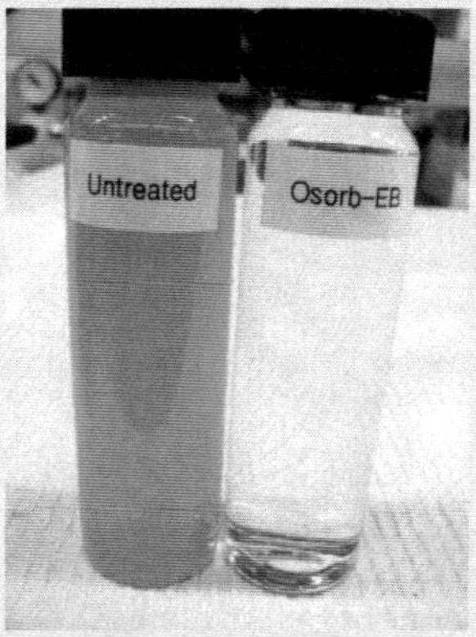

TOC before = 265 ppm
TOC after = no detect
0.4%*w/v* Osorb

Surfactant Alcohol Removal

Goal is to improve performance on hard to treat waters and remove highly soluble chemicals used in fracking.

2-butoxyethanol
50 ppm

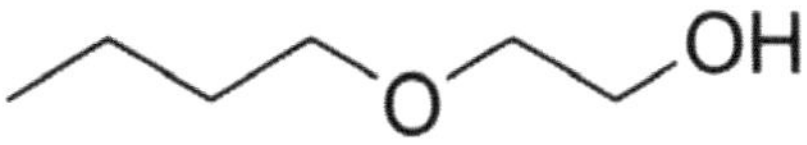

Osorb Type	Percent Extraction
Regular (1%*w/v*)	**42%**
Osorb-EB (0.3%*w/v*)	**66%**

PW Recently Fracked Well (Ohio)

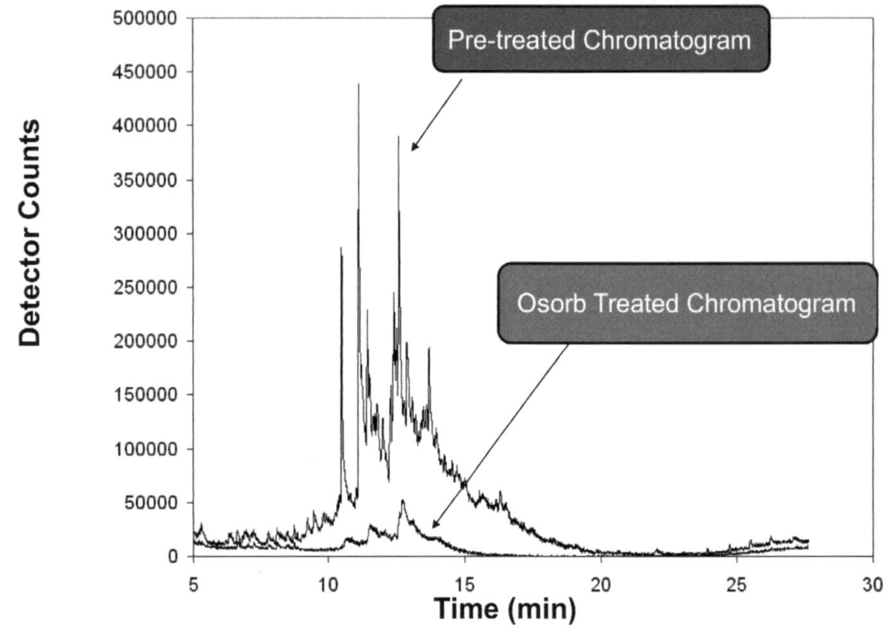

Penn/WV Sample Late Stage Flow-Back Water

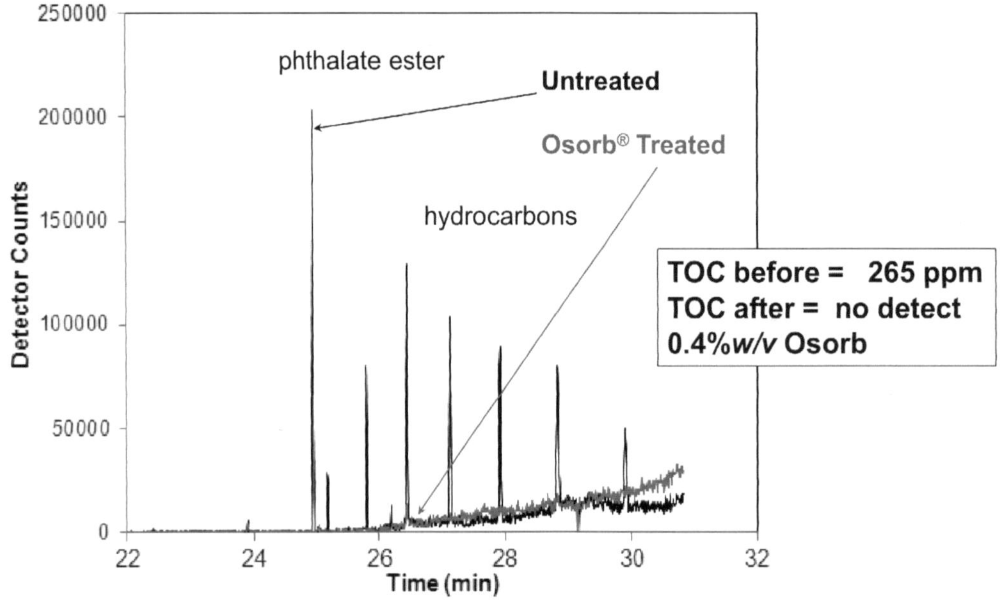

Flow Back Water Testing: Skid Mounted Unit

Collaboration with Texas A&M GPRI

Eagle Ford Play: Fall 2010

Analyte	Untreated Water (ppm)	25 gal (ppm)	50 gal (ppm)	75 gal (ppm)	100 gal (ppm)	% Reduction
Benzene	4.24	0.114	0.139	0.194	0.206	95.14%
Ethylbenzene	0.094	0.0016	0.0022	0.0025	0.0027	97.12%
Toluene	0.244	0.046	0.55	0.088	0.094	61.47%
1,2,4-Trimethylbenzene	0.01	0.0011	0.0014	0.0019	0.0019	61.47%
Total Xylenes	0.062	0.011	0.015	0.018	0.019	69.35%
Oil&Grease	11.5				0	100.00%

Flow back water supplied by DOE, Lab Tests

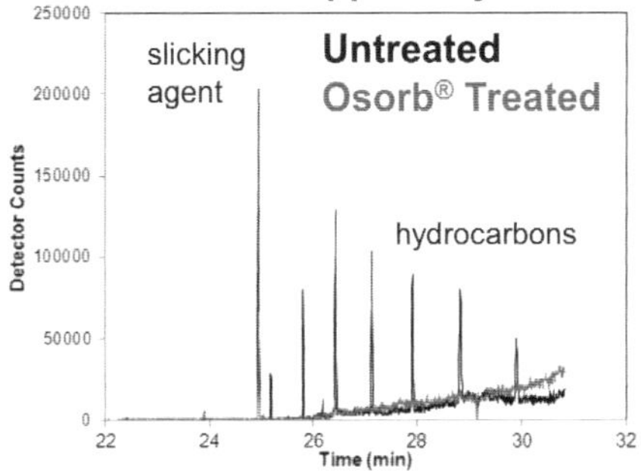

4 gal/min skid mounted cartridge unit: Osorb removes 80-100% of BTEX, oil, broken polymers, and slicking agents

60 gal/min Water Treatment System

Funding from National Science Foundation and U.S. Department of Energy for pilot scale testing in the field, produced water and flow back

Trailer and Skid-Mounted Systems Available (4-60 gal/min)
Skid system tested by Texas A&M University

NSF U.S. DEPARTMENT OF ENERGY ABS MATERIALS PWA

No Impetus to Treat Flow Back Water?

• Biggest competition to water treatment technologies are injection wells and blending.

• Understanding on to blend and reuse flowback water is allowing some operators to lessen fresh water needs. Little need for water treatment.

• Low cost of natural gas makes expenditures on water treatment even less attractive.

• Government regulation on water treatment and controls is limited.

Pilot Scale Testing

• 60 gal/min treatment system and 4 gal/min cartridge unit were built using DOE SBIR Phase II funding.

•Pilot testing available now.

Costs

- Cost. Disposal well costs run $0.75-$1.75/bbl. Hauling costs $0-$5/bbl.
- Our treatment costs run $3.50-$7/bbl.

We want to drive down our costs by making our largest cost center, regeneration easier, faster and cheaper.

Need data from users.

Rare Earth Mining from Shale Gas Water

Rare earths elements are not rare, but formations of high concentration are hard to find.

Found in alluvial deposits where freshwater meets salt water.

Ideal location would be in ancient estuary environments.

Many are buried in shale deposits.

Hydraulic fracking is exploring deep shale deposits.

Utica shale shows regions where rare earth element concentrations are in excess of 4,000 ppm.

Exploring synergistic extraction of REEs and hydrocarbons in PW

Rare Earth Extraction from Shale Gas Water

Challenge is extracting REE from Group II cations.

Creating a type of Osorb that duplicates the multistage liquid-liquid extraction process use in conventional hydrometallurgical processes in a single core-shell particle

Goals:

1) **Rapid sampling system using hand-held XRF**
2) **Larger scale extraction system for PW.**

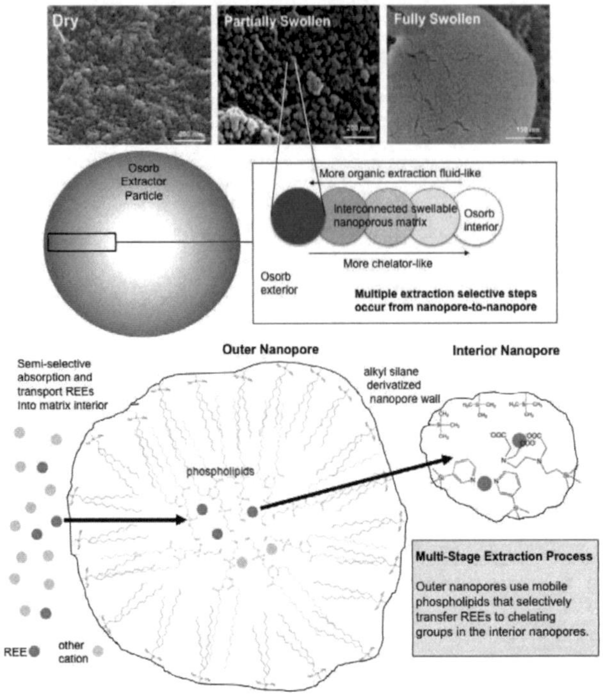

Looking for samples from operators

- DOD funding to measure flow back water and produced water for high value rare earth elements. Samples needed. Confidential. High need to U.S. interests.
- We are creating test kits so operators can screen their own produced water to prospect for rare earths to find additional high value products in their wells.

Funding Sources

• NSF-CAREER Award #0238808, 2003-2008, Analytical and Surface Chemistry. TNT sensors. First observation of Osorb Spring 2004.

• NSF Chemical and Biological Separations Division Award #0939371 2008-2011. Application of Osorb towards water treatment including first pilots.

•NSF RAPID 2010: Gulf of Mexico Clean-Up.

•Two NSF-SBIR Phase II (including Phase IIB). Produced water and Stormwater treatment.

•Department of Energy SBIR-Phase II: Flow back water treatment Current

•Two ARPA-E proposals pending. Methane and rare earth element extraction.

For more information...

Contact info:
Paul L. Edmiston, CSO
p.edmiston@absmaterials.com

Stephen Spoonamore, CEO
s.spoonamore@absmaterials.com

www.absmaterials.com

330-234-7999

Edmiston, P. L.; Underwood, L. A. Absorption of Dissolved Organic Species from Water Using Organically Modified Silica that Swells. *Separation and Purification Technology* 66, 532-540 (2009).

Burkett, C. M.*; Underwood, L. A.*, Volzer, R. S.*; Baughman, J. A.*; Edmiston, P. L. Organic-Inorganic Hybrid Materials that Rapidly Swell in Non-Polar Liquids: Nanoscale Morphology and Swelling Mechanism. *Chemistry of Materials* 20, 1312-1321 (2008).

Burkett, C. M.; Edmiston P. L.; Highly Swellable Sol-Gels Prepared by Chemical Modification of Silanol Groups Prior to Drying. *J Non-Crystalline Solids ,351* , 3174-3178 (2005).

Edmiston, P.L.; Campbell, D.P.; Gottfried, D.S.; Baughman, J.*; Timmers, M.M.* Detection of Trinitrotoluene in the Parts-per-Trillion Range Using Waveguide Interferometry, *Sensor & Actuators B.* 143, 574-582 (2010).

Acknowledgements and References

Support:

National Science Foundation
U.S. Department of Energy
Ohio EPA

Collaborators:

Dr. Hanbae Yang
Dr. Tatiana Eliseeva
Dr. Stephen Jolly
Justin Keener

Students:

Zachary Harvey
Alison Chin
Noel Mellor
Christine Kasprisin
Melissa Morgan
Paige Piper

Edmiston, P. L.; Underwood, L. A. Absorption of Dissolved Organic Species from Water Using Organically Modified Silica that Swells. *Separation and Purification Technology* 66, 532-540 (2009).

Burkett, C. M.*; Underwood, L. A.*, Volzer, R. S.*; Baughman, J. A.*; Edmiston, P. L. Organic-Inorganic Hybrid Materials that Rapidly Swell in Non-Polar Liquids: Nanoscale Morphology and Swelling Mechanism. *Chemistry of Materials* 20, 1312-1321 (2008).

Burkett, C. M.; Edmiston P. L.; Highly Swellable Sol-Gels Prepared by Chemical Modification of Silanol Groups Prior to Drying. *J Non-Crystalline Solids ,351* , 3174-3178 (2005).

Edmiston, P.L.; Campbell, D.P.; Gottfried, D.S.; Baughman, J.*; Timmers, M.M.* Detection of Trinitrotoluene in the Parts-per-Trillion Range Using Waveguide Interferometry, *Sensor & Actuators B.* 143, 574-582 (2010).

www.absmaterials.com
pedmiston@wooster.edu 330-234-7999

Unconventional Gas Wastewater Treatment – Perspectives from a Process Provider

Harish R. Acharya, Ph.D.
GE Global Research Center, Niskayuna, NY

Mark R. Wilson
GE Power & Water, Woodlands, TX

November 1, 2012
AIChE-A&WMA Joint Workshop on Shale Oil & Gas E&P – Water Challenges and Opportunities
Pittsburgh, PA

GE Mobile UF/RO

GE Mobile Evaporator

imagination at work

1 /

Outline

- Scope, Benefits & Challenges
- Frac flowback & produced water characteristics – variation
- Key Contaminants
- "Product water" requirements
- Treatment process options
- Economic considerations
- Challenges – technical and economic
- Conclusions

imagination at work

2 /

Scenarios for Flowback Wastewater Treatment

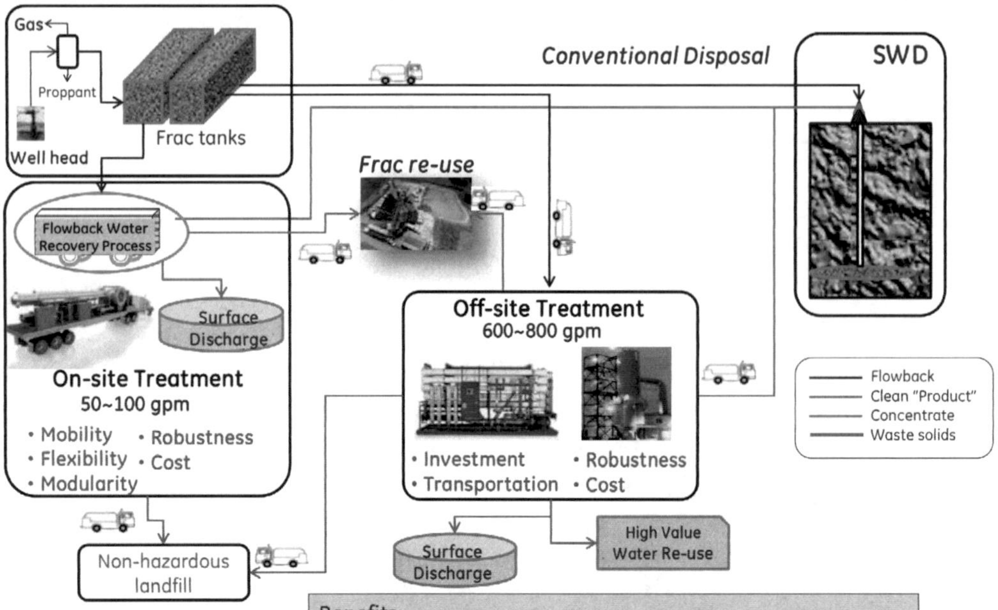

Benefits:

- Reduced net fresh water used in hydrofracturing process
- Reduced wastewater and associated disposal costs
- Recover produced water for surface (agricultural) use

3

Frac Flowback: TDS ranges for various shales

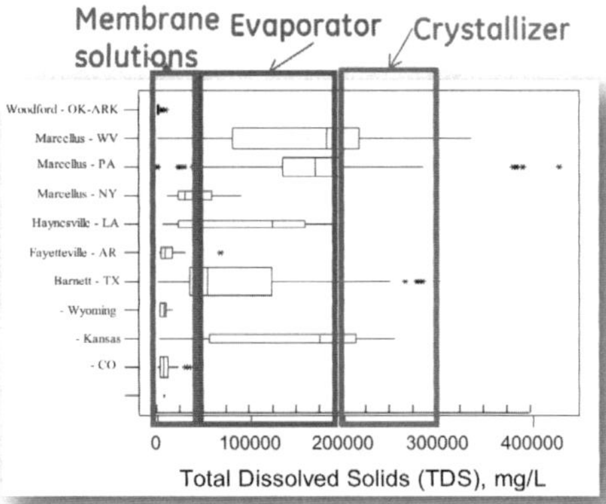

Shale Play	Feed TDS range (Average), mg/L	Most common flowback handling process	Desal process recommended
Western shales	5K ~ 15K	Surface discharge or SWDisposal	Reverse Osmosis
Fayetteville	10K ~ 30K (20K)	Clarify & re-use or SWDisposal	Reverse Osmosis
Anadarko	20K ~ 30K	SWDisposal	Reverse Osmosis
Woodford	30K ~ 120K (40K)	SWDisposal	RO or Evaporator
Barnett	50K ~ 180K	SWDisposal	Evaporator
Haynesville	>150K	SWDisposal	Evaporator
Marcellus	70K - 250K (100K)	Clarify & re-use or SWDisposal	Evaporator/Crystallizer
Utica	150K~200K	Clarify & re-use or SWDisposal	Evaporator/Crystallizer
Bakken	>250K+	SWDisposal	Crystallizer

4/

Frac Flowback Characteristics

Frac Water Usage

Dependent upon:

- Type of well
- Geographical location
- Operator bias/ experience

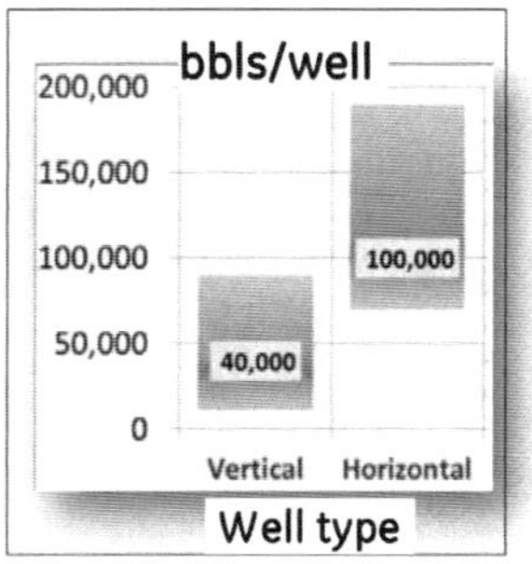

Frac Flowback

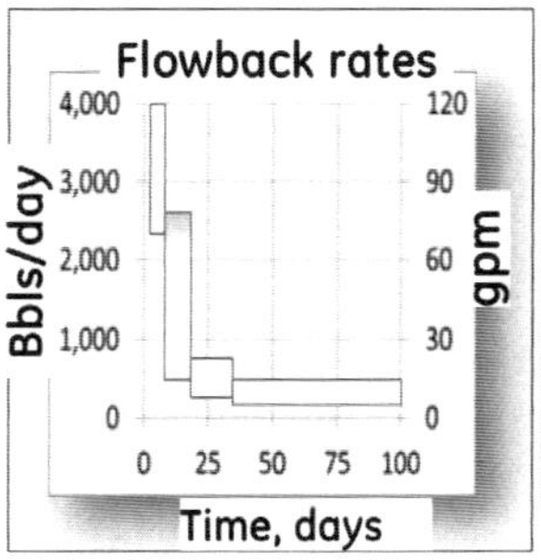

Time, days	Flowback recovery, % frac fluid
1 - 5	10 ~ 25%
5 - 15	8 ~ 12%
15 - 30	1 ~ 5%
30 - 90	1 ~ 2%
Total	20 ~ 40%

Examples (Woodford shale)

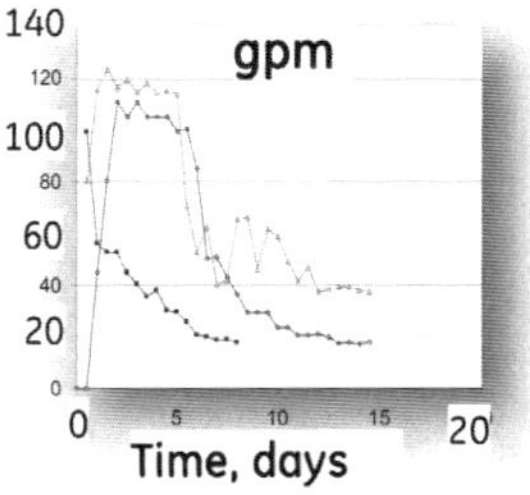

Flow rates:
- Decrease with time
- Profiles differ by well

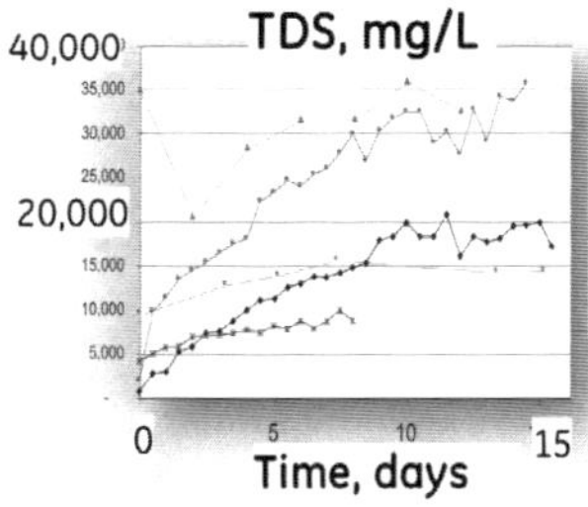

TDS:
- Increases with time
- Profiles differ by well

Caveat: "Every well is different"

imagination at work

5 /

Frac Flowback: Key contaminants

Impact for re-use in hydrofracturing

Particulates Suspended solids	Plugging
Oil and grease Dissolved Organics Volatile Organics Total Dissolved Solids (TDS) Chlorides	Fluid stability
Iron Hardness (Ca, Mg) Barium Strontium Silica Sulfates	Scaling
Biological counts	Bacterial growth
NORM (Normally Occurring Radioactive Materials)	Radioactivity

imagination at work

6 /

Frac Flowback Water Sampling

Woodford Shale Site Days 1-14

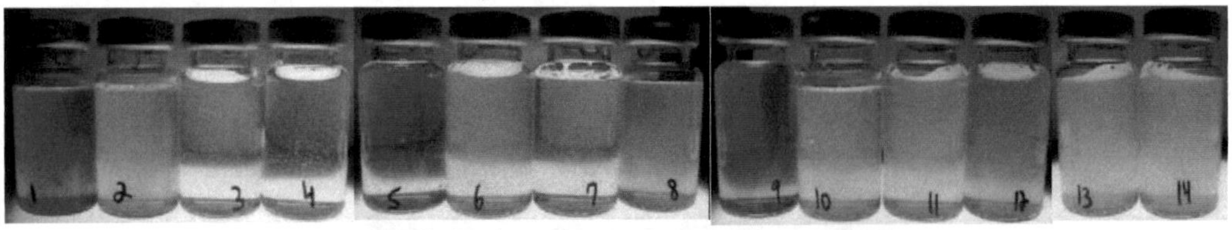

Marcellus Shale Site Days 1-12

imagination at work

Appearance & composition changes with time

7 /

Flowback samples: Contaminant profiles

Woodford shale: Site-1 (Days 1-14) & Site-2 (Day 26)

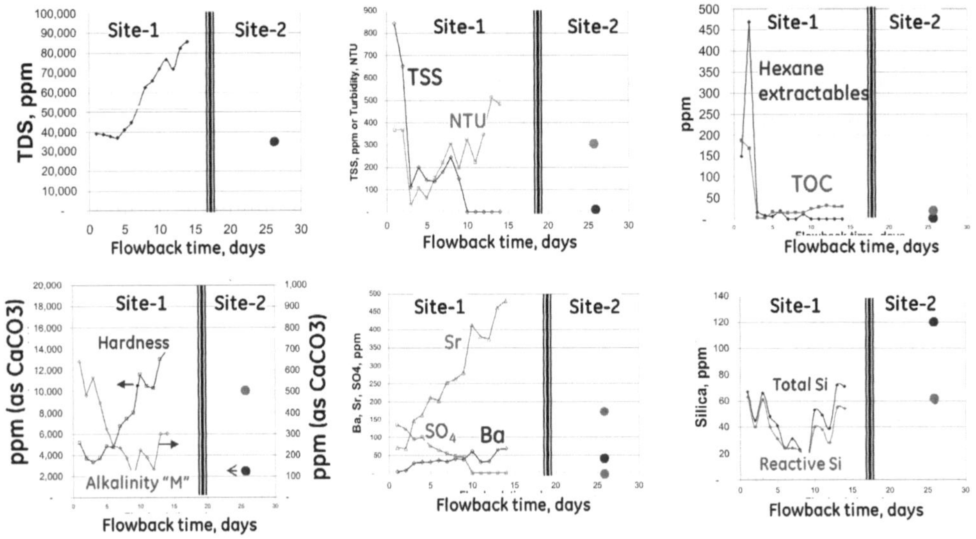

imagination at work

Challenge: Composition highly variable with time
......and location - geology, operating conditions

8 /

Frac Flowback Recovery Process

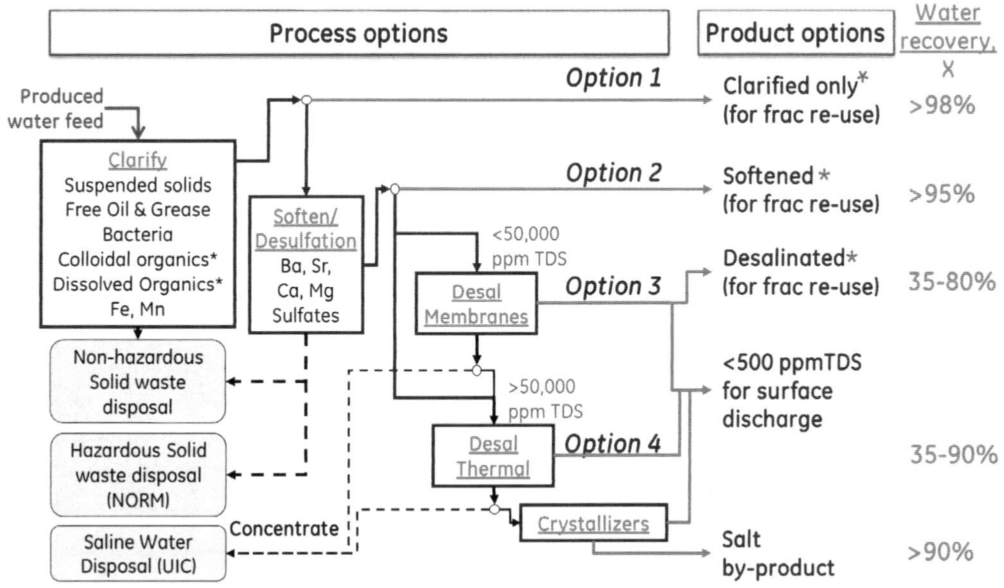

Cost, Recovery, Robustness, "Green"

imagination at work

*Issue: High variability in "product" specs

9 /

Cleanup Technology Alternatives: Rating for Key Criteria

Contaminant	Technology
Solids/Debris	Filtration
Suspended Solids	Coagulation/Floc MicroFiltration
Oils & organics	Flotation Memb Bio Reactor Oxidation
Hardness	Chemical Softening Nanofiltration
Desalination	RO, FO, Evaporators

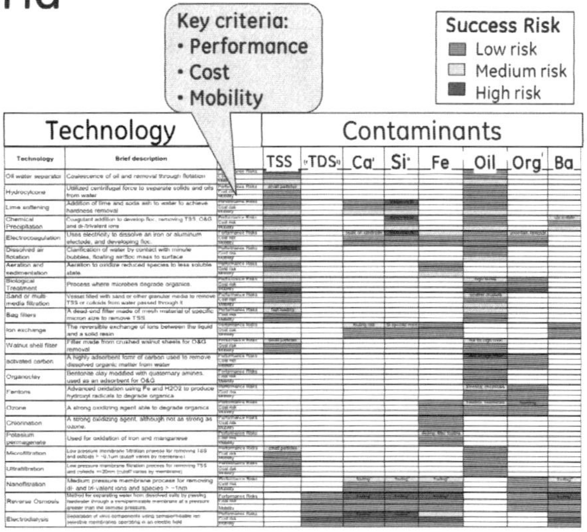

• Wide spectrum of water treatment technologies available
• Process choice highly dependent on water chemistry, cost, performance & robustness criteria for particular shale location

10 /

On-site Mobile Rig: System cost analysis

Flowback Water Recovery Process

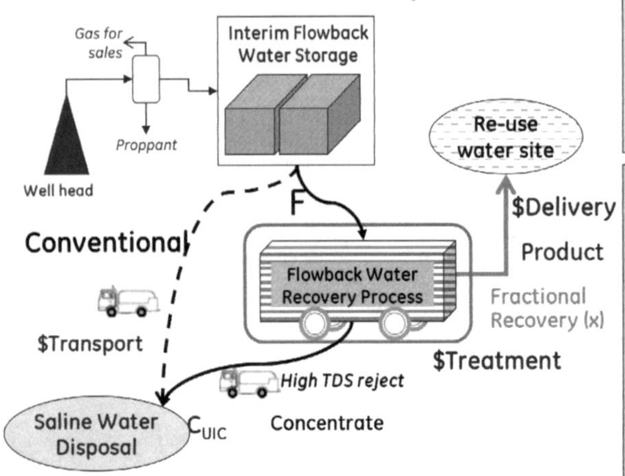

Conventional Disposal $/bbl frac water

\$Conventional = F \$Disposal

\$Disposal = \$Transport + \$UIC \$/bbl disposed water

FWRP $/bbl frac water

\$FWRP = Product*\$Product + Concentrate* \$Disposal

\$FWRP = x F \$Product + (1-x) F \$Disposal

X = Product/F, Fractional water recovery

\$Product = \$Treatment + \$Delivery - \$Fresh Water

\$/bbl Product Water

Key objective:

Minimize \$FWRP vs. \$Conventional costs for given local \$Disposal (Transport + UIC) costs by:

- ⬆ X, fractional water recovery
- ⬇ \$Product (\$Treatment Process + \$Delivery) costs

Water recovery via desalination

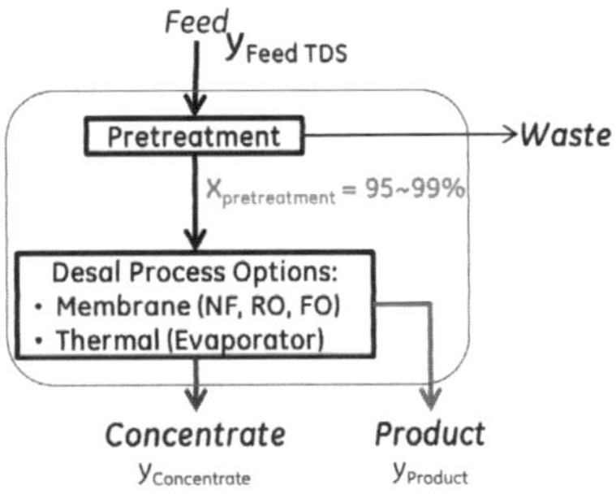

$$\text{Recovery, X} = \frac{\text{Product rate}}{\text{Feedrate}}$$

Limitations to Concentrate TDS :

- RO, FO: Water chemistry, osmotic pressure & membrane fouling
- Evaporator: tube fouling due to precipitation & corrosion (MOC)

Water Recovery vs. Concentrate TDS for given Feed TDS

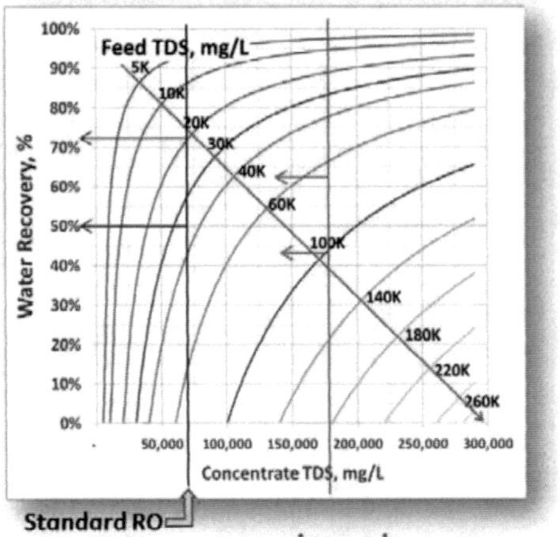

Standard RO
yConcentrate = 70K

Examples:

- Seawater (35K feed) ➔ X = 50%
- Fayetteville (20K feed) ➔ X = >70%

Standard Evaporator, FO, etc.
yConcentrate = 150K ~ 200K

Examples:

- Barnett (70K feed) ➔ X = 62%
- Marcellus (100K feed) ➔ X = 43%

imagination at work

Water recovery via desalination: Technology impact

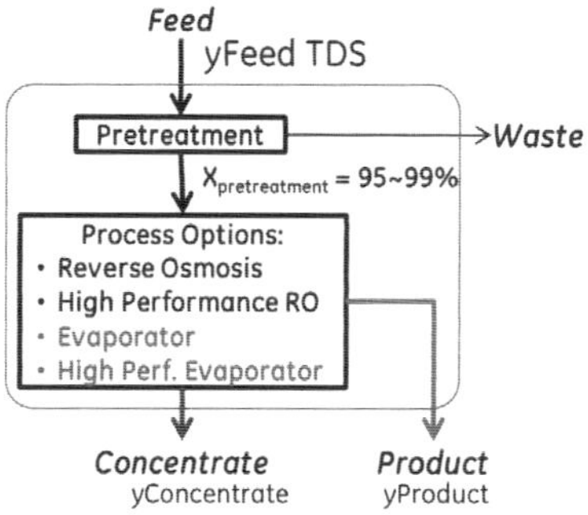

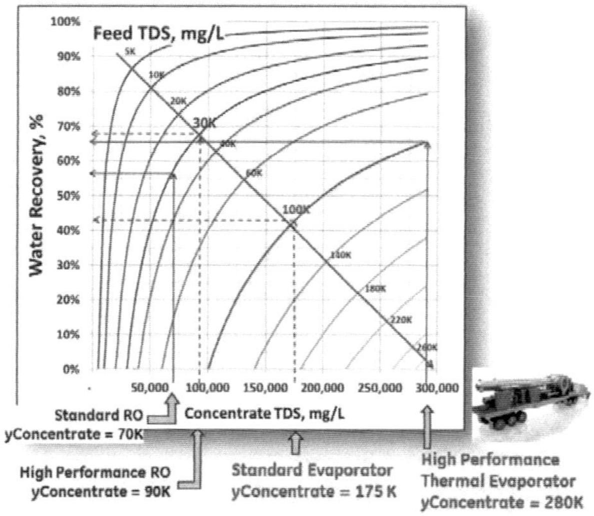

Case A: Low TDS Feed
yFeed = 30K mg/L
- Standard RO, X = 56%
- High Perf. RO, X = 67%

Case B: High TDS Feed
yFeed = 100K mg/L
- Standard Evaporator, X = 43%
- High Perf. Evaporator, X = 65%

Need to optimize recovery vs. capital/energy/labor costs

13 /

Water recovery via desal: Applicability to UG plays

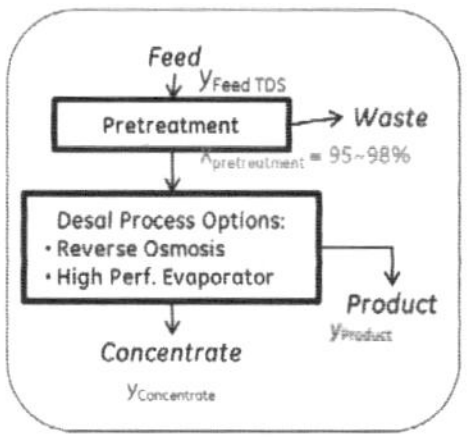

$$\text{Recovery, } X = \frac{\text{Product rate}}{\text{Feedrate}}$$

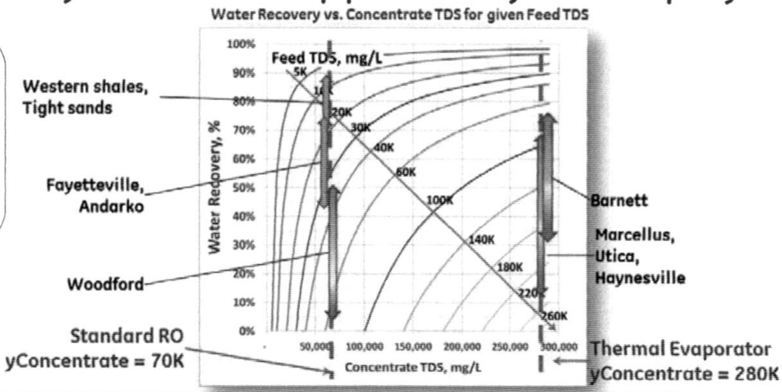

Shale Play	Feed TDS range (Average), mg/L	Most common process	Desal process recommended	Anticipated water recoveries, %
Western shales	5K ~ 15K	Surface discharge or SWDisposal	RO	95% - 75%
Fayetteville	20K ~ 30K	Clarify or SWDisposal	RO	75% - 55%
Andarko	20K ~ 30K	SWDisposal	RO	75% - 35%
Woodford	30K ~ 80K	SWDisposal	RO or Thermal	60% - 30%
Marcellus	70K - 250K (100K)	Clarify & re-use	Evaporator	70%~ %
Utica	150K~200K	Clarify & re-use	Evaporator	45% - 30%
Barnett	70K ~ 180K	SWDisposal	Evaporator	75% - 35%
Woodford	30K ~ 80K	SWDisposal	RO or Thermal	60% - 30%
Haynesville	150K	SWDisposal	Evaporator	45%
Bakken	250K+	SWDisposal	Crystallizer	90-95%

14 /

On-site Mobile Rig: Potential costs

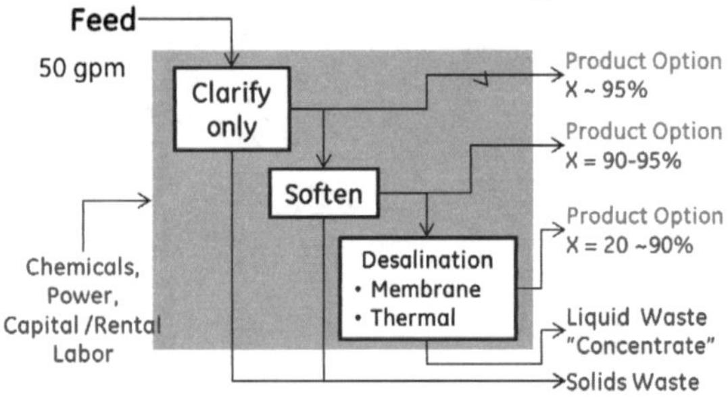

$FWRP = $Recovery + $Conc_Disposal $/bbl feed

- $Recovery = $Process + X* ($Product Delivery - $Product Value)
- $Conc_Disposal = $Disposal*(1-X)
- $Disposal = $Transportation + $SalineWD
- Recovery, X = Product/Feed

$Conventional = $Disposal $/bbl feed

%Savings = 1 - $FWRP/$Conventional

Potential cost ranges for 50 gpm systems*

Process	Setup:	Varies ~ $5-$10K/setup
	Capital/ Rental:	$500 - $2,000/day ($0.23 - $0.90/bbl feed)
	Energy:	$0.05 - $0.50/bbl feed
	Labor:	$0 - $1.50/bbl feed
	Chemicals :	$0.05 - $0.50/bbl feed
	Solid Waste :	$0.05 - $0.50/bbl feed
	Total:	**$1 - $5/bbl feed**

Product Delivery : $0-$3/bbl product

Product Value : $0-$1/bbl product

Liquid Concentrate Disposal:
$1-$20/bbl concentrate

*Caveat: Costs highly dependent on shale location, feed composition, "product" specs.
Costs per bbl feed will be lower for larger capacities
Cost burden of regulations not considered

imagination at work

15 /
Harish R. Acharya, Global Research

Cost sensitivity to key parameters

On-site Mobile Rig

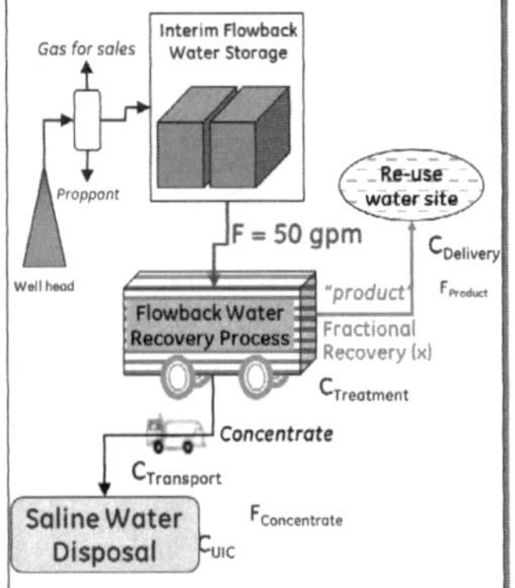

$Disposal = $Transport + $UIC
$FWRP = $Process
+ X ($Product delivery- $Product Value)
+ (1-X) $Disposal

Example: Cost estimation for 50 gpm (1,700 bbl/day) system (24 hrs x 30 days) vs. Conventional Disposal @$7/bbl

Key variables: Capex cost
%Recovery, X

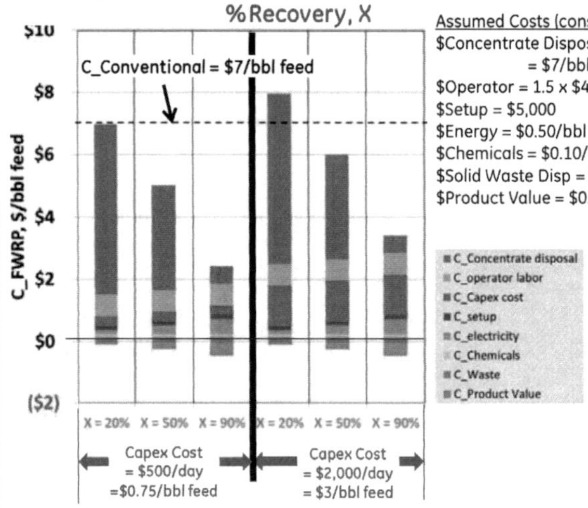

Assumed Costs (constant):
$Concentrate Disposal = $7/bbl concentrate
$Operator = 1.5 x $40/hr
$Setup = $5,000
$Energy = $0.50/bbl product
$Chemicals = $0.10/bbl feed
$Solid Waste Disp = $0.15/bbl feed
$Product Value = $0.50/bbl delivered

Key parameters for small-scale flowback water applications: Water recovery, Equipment costs, Operating Labor & Energy

16 /
GE Global Research/GE Water

On-site Mobile unit: Economic benefits

Savings = 1- \$FWRP/\$Conventional

Example

Assume: $C_{Prod\ Disposal} - C_{Prod\ Value} = \sim 0$
(e.g., "surface discharge for agri use"

Breakeven point (Savings = 0)

Case	\$Process \$/bbl	X, %	\$Disposal \$/bbl
A	\$2	90	\$2.20
B	\$2	50	\$4.00
C	\$4	90	\$4.50
D	\$4	50	\$8.00

Savings sensitivity to local disposal costs for key operating parameters: \$Process & Recovery, X

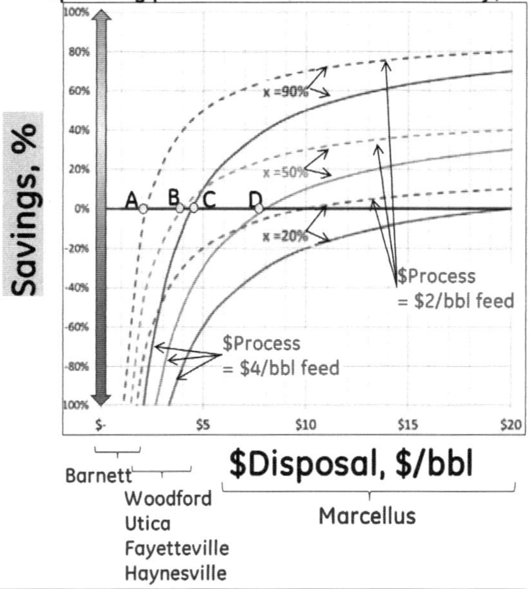

Economic benefits dependent on shale play, product specs, volume processed, choice of separation technology & operating parameters, regulations/incentives

Frac flowback/produced water – Technology issues

Status:

- Water treatment technologies are mostly commercially available from industrial wastewater applications.

Innovation Challenges for the flowback/produced water market:

There is urgent need to:

- Increase % water recovery while reducing process operating costs
- Decrease Capital Equipment (Lifetime) costs
- Decrease footprint for mobility
- Increase robustness to highly variable flow rates and compositions, ...
  ...and surprises – slugs of oil & grease, solids, volatile hydrocarbons
- Increase remote sensing & control capabilities to reduce operating labor costs

Frac flowback/produced water – Economic issues

Status:

Overall economic feasibility is highly dependent on shale play specifics, water chemistry, product specs, local alternative disposal costs, and local regulations.

Challenges for a water treatment process provider:

- Changing economics scenarios: e.g. increased re-use after minimal treatment by blending with fresh water (Marcellus, Utica).
- Low demand/attractiveness in many shale plays (Barnett, Haynesville, Bakken, Woodford) due to inherently high salinity of flowback water and/or easy availability of UIC disposal sites
- "Product" specs: High variability of re-use water quality requirements across industry.
- Changing demand: Need for water recovery in many shale plays is weather (drought) dependent.
- Uneven regulations: Surface discharge (for non-potable use) regulations vary by localities – who bears responsibility for the treated water?

imagination at work

19 /

Summary and Conclusions

- Flowback water is not a uniform "raw material" from a process development perspective.
  - Composition highly variable – function of geology, operating conditions & life of well
  - Flow rates initially high (~ 100 bbl/hr) but decreases to ~ 100 bbl/day over 30~60 days.
- Product quality requirements highly dependent on end-use ("Clarify", "Soften" or "Extent of Desalination")
- Water treatment technologies are mostly commercially available from industrial wastewater applications. But need to adapt for increased robustness and decreased costs & footprint (for mobility).
- Overall economic feasibility is highly dependent on shale play specifics, water chemistry, product specs, local alternative disposal costs, and local regulations.

imagination at work

20 /

Thank you!

ACKNOWLEDGEMENT

Part of this material is based upon work supported by the US Department of Energy, National Energy Technology Laboratory under Award Number DE-FE0000784.

imagination at work

AIChE/A&WMA Joint Workshop
Shale Oil & Gas E&P-
Water Challenges and Opportunities
November 2, 2012

David B. Burnett
Department of Petroleum Engineering GPRI
Texas A&M University

Harold Vance Department of PETROLEUM ENGINEERING TEXAS A&M UNIVERSITY

GPRI

How Engineers (and geologists) Can Save America

EFD

There is Plenty of Water in the World – 70% of the Earth's Surface, Remember?

U.S. Future Natural Gas Production from Unconventional Resources

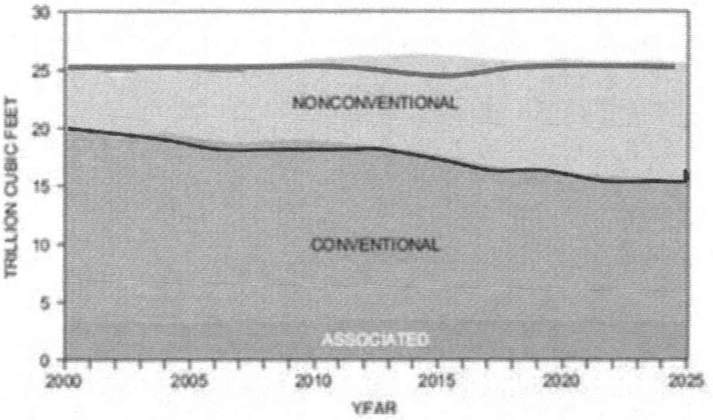

U.S. Lower 48 and Non-Arctic Canadian Gas Production by Type

How Innovative Are We?

Can We Meet the World's Energy Demands for the Next 100 Years?

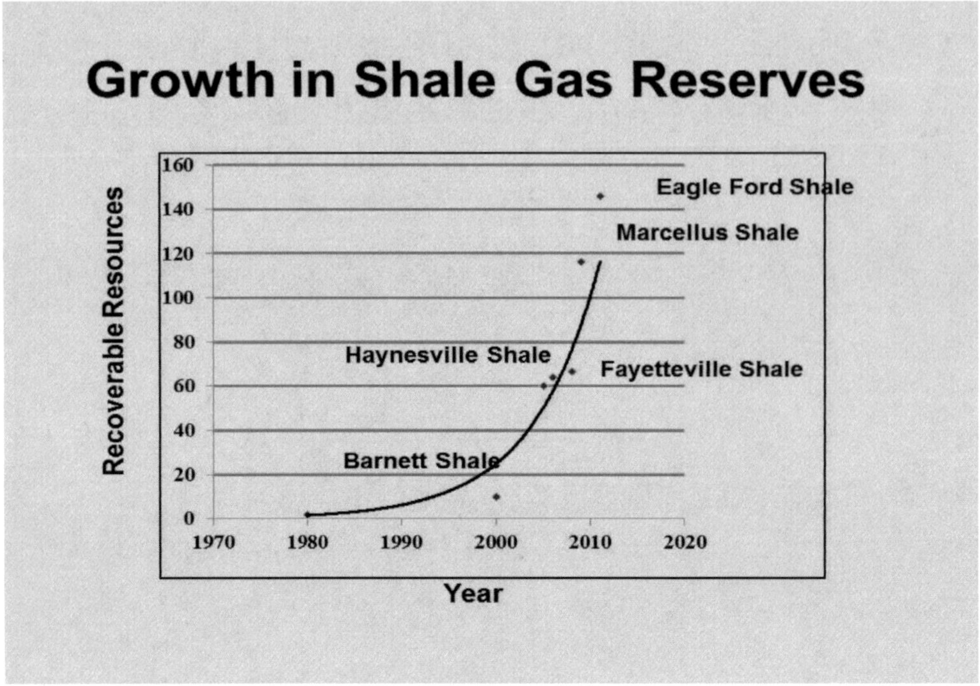
Growth in Shale Gas Reserves
Recoverable Resources
160
140
120
100
80
60
40
20
0
1970
1980
1990
2000
2010
2020
Year
Eagle Ford Shale
Marcellus Shale
Haynesville Shale
Fayetteville Shale
Barnett Shale

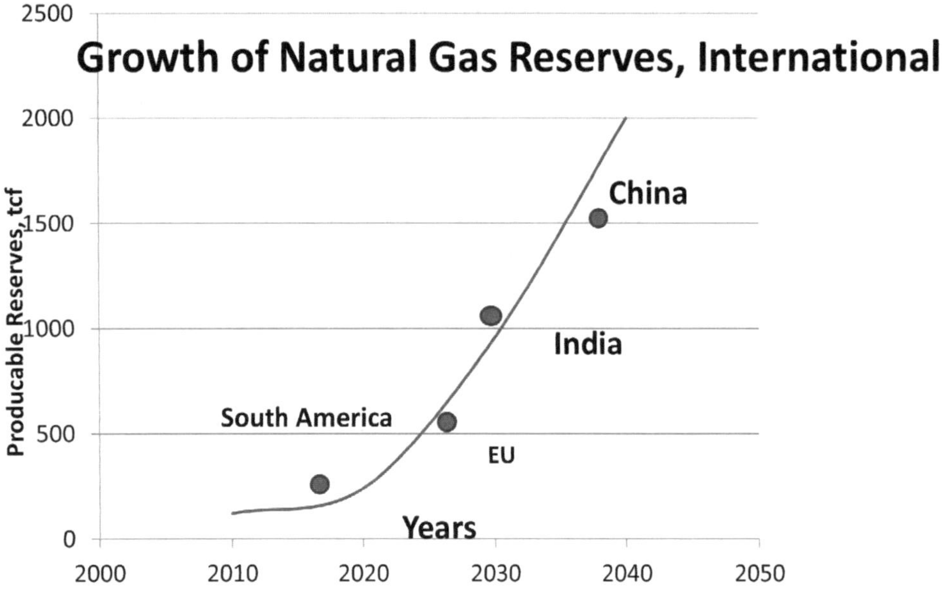
Growth of Natural Gas Reserves, International
Producable Reserves, tcf
2500
2000
1500
1000
500
0
2000
2010
2020
2030
2040
2050
Years
China
India
South America
EU

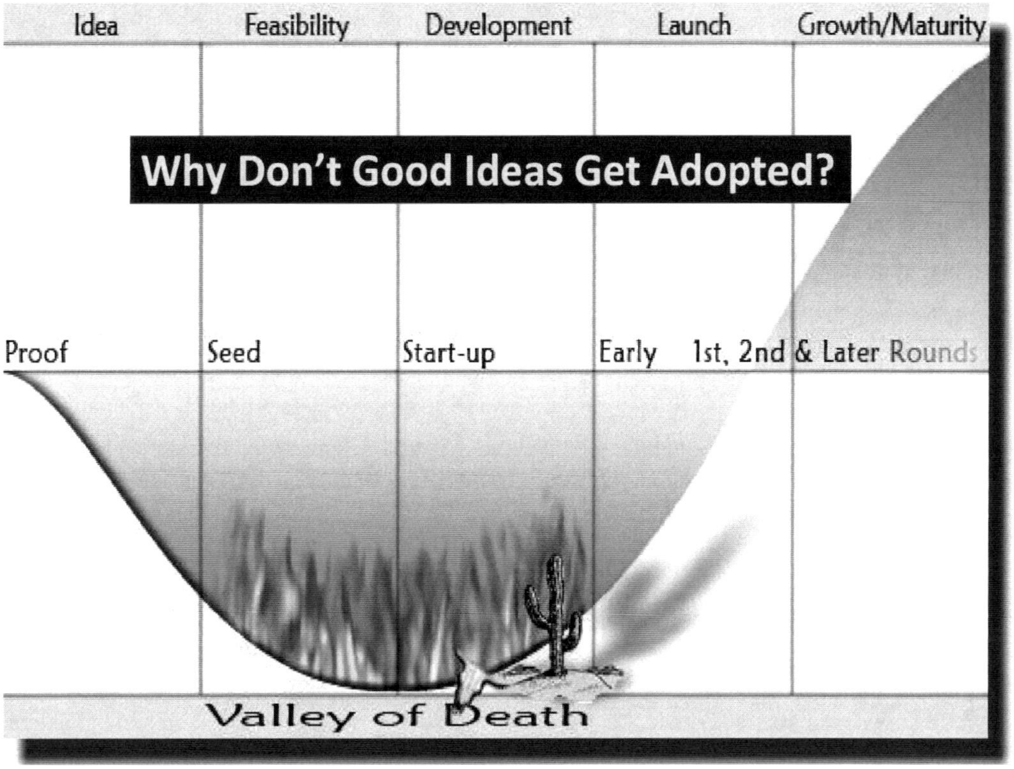

What are some Ways to Bridge the Gap and Accelerate Innovation?

The Environmentally Friendly Drilling Systems Program

Interstate Oil & Gas Compact Commission
Awarded October 5, 2009
CHAIRMAN'S STEWARDSHIP AWARDS

Demonstrating technologies that reduce environmental footprints

WWW.efd-aas.org

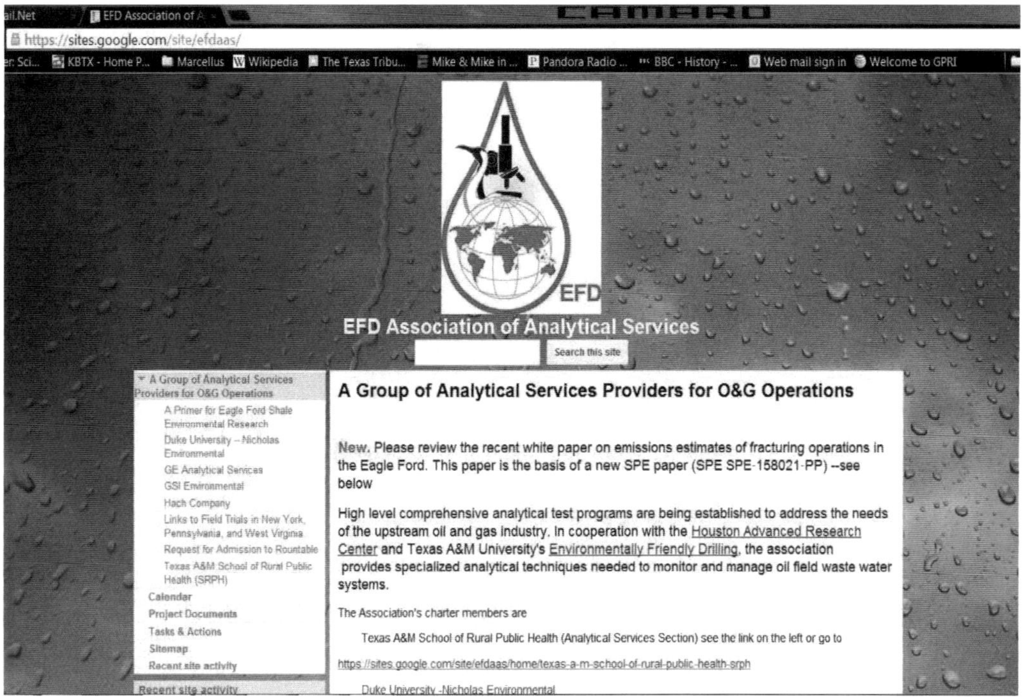

Barriers to Success; Overcoming Barriers

- Technical Issues
  - Technology translation from pilot plant to field ops
  - Scheduling/priorities
  - Financial challenges
  - Data Gathering –
    - dissemination
    - Confidentiality of Information
- `Regulatory Issues/Problems
  - Delay in Permitting
  - Communication Lapses
  - Overlap of Authority
  - Inability to Stay Current with Technology
  - Staffing, public and private
- Community Dialog
  - Recognizing that "Traditional Knowledge" exists
  - Recognizing the impact of extremist views

Field Trials
- Site Candidates: Western and Eastern

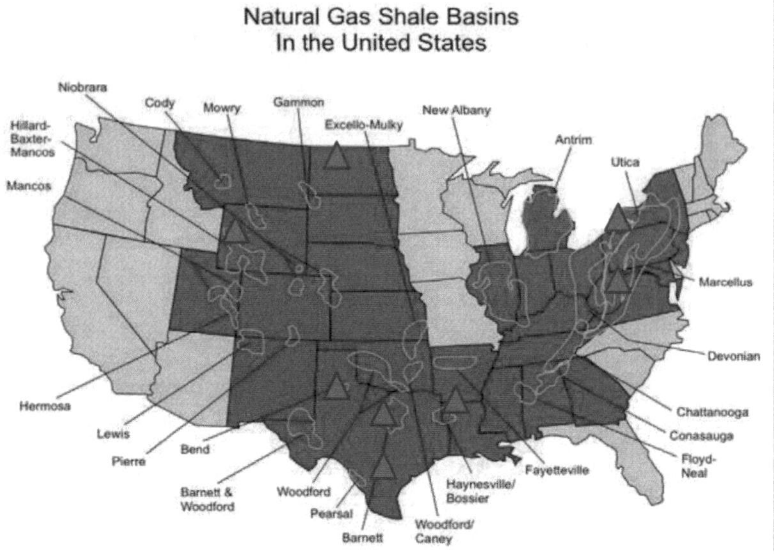

Mobile Testing Laboratory to Support Field Trials

- Running Field Trials

Figure 2A shows a satellite view of the mobile unit in Chenango Co. NY

Figure 2B shows the mobile laboratory on its environmental protection apron at the field site. The large black poly tank contains waste fluids

How a Trial is Scheduled – Timeline and Tasking

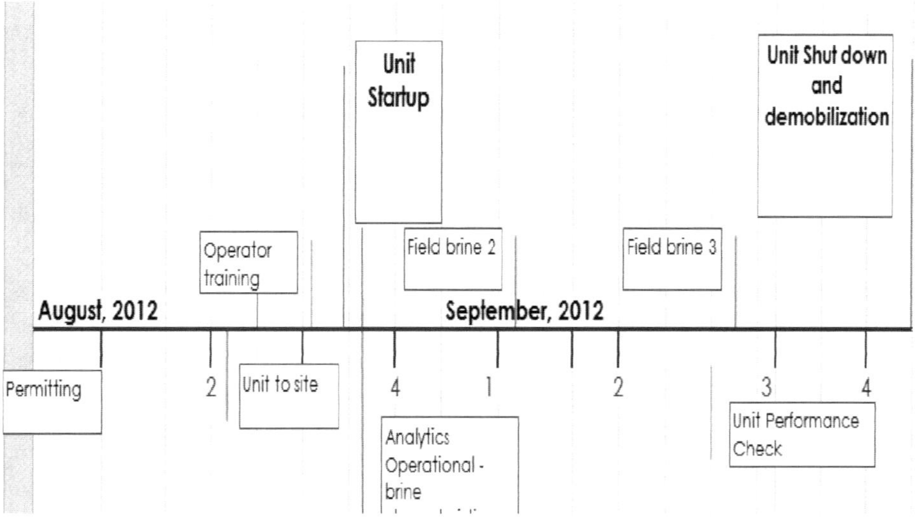

Process Train Schematic

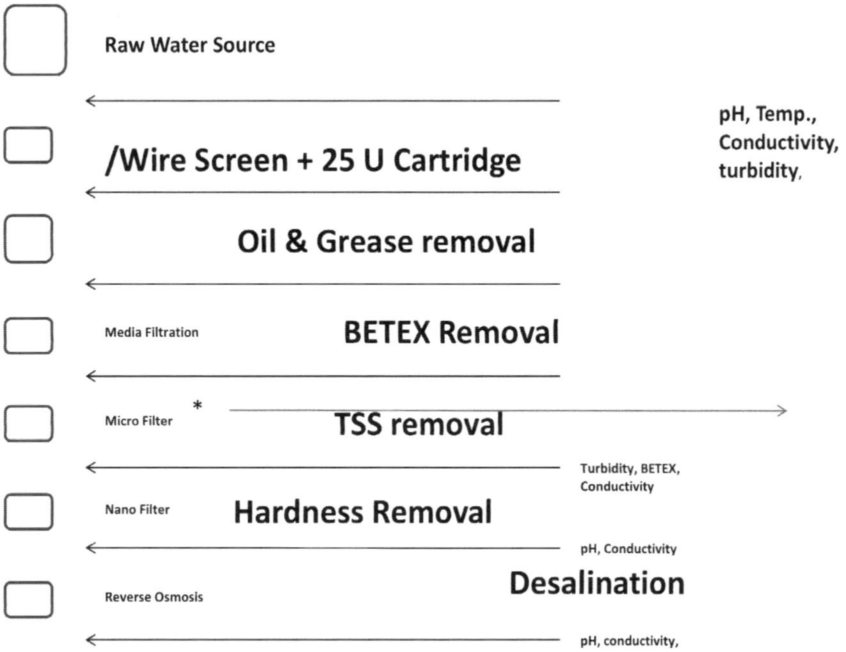

Identifying Key Technology Components

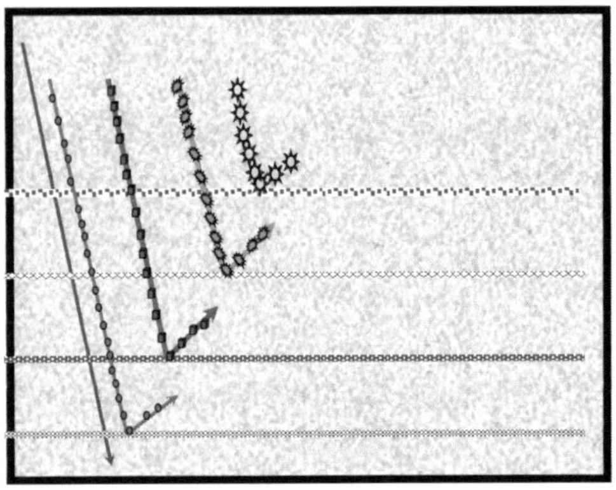

Micro Filtration (MF) (10-0.1μm)
Bacteria, suspended particles

Ultrafiltration (UF) (0.05-0.005μm)
Colloids, macromolecules

Nanofiltration (NF)
Sugars, dyes, divalent salt ppts.

Reverse Osmosis (RO)
Monovalent salts, ionic metals

Floor Plan Layout for Mobile Trailer and Water Tanker (2010)

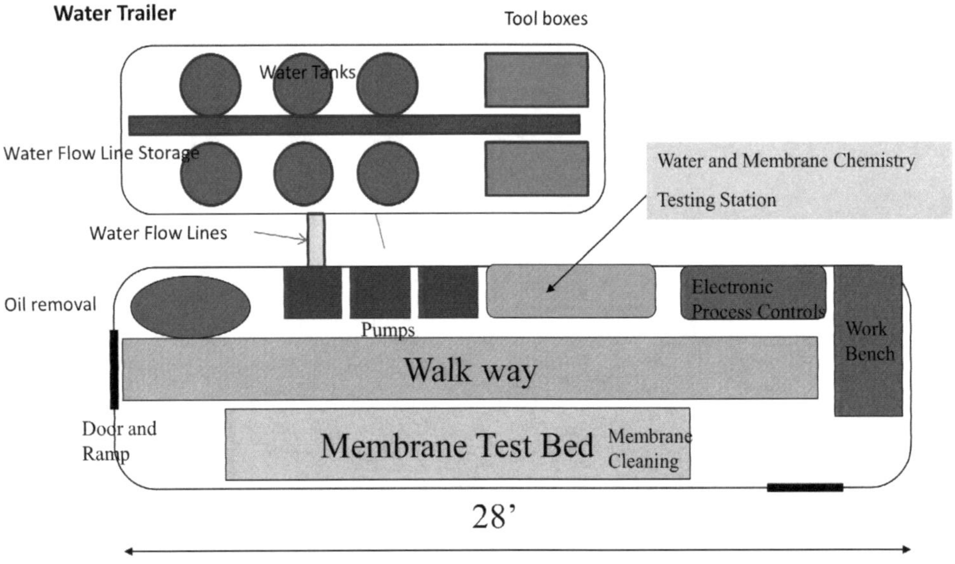

Pennsylvania Frac Pond is Aerated to Avoid Souring

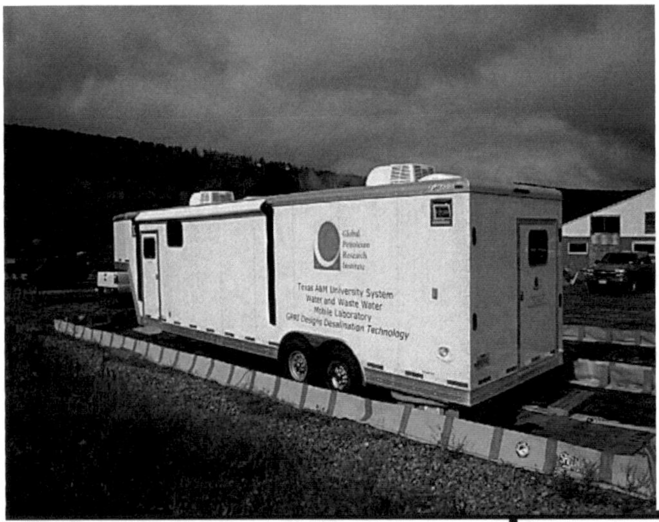

Texas A&M Marcellus Shale Brine Re-Use

Deliverables

- Identify optimal treatment techniques for ultra-high brine concentrations
- Provide information to community leaders

Status

- ***Current field trials in New York***
- ***A&M Separation Sciences pilot testing new technology***
- ***Creation of new Analytical Services Roundtable***

New York Field Trial of Ultra-High Salinity Brine Pre-treatment: Texas A&M Environmentally Friendly Drilling Technology for the Marcellus Shale –SPE 158396

Low Pressure Media Filtration

Oil and Grease Removal

High Pressure Membranes and On-Line Sensors

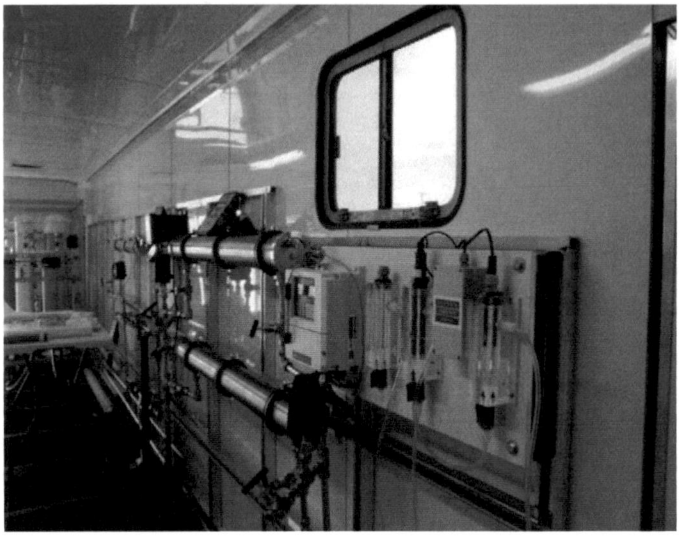

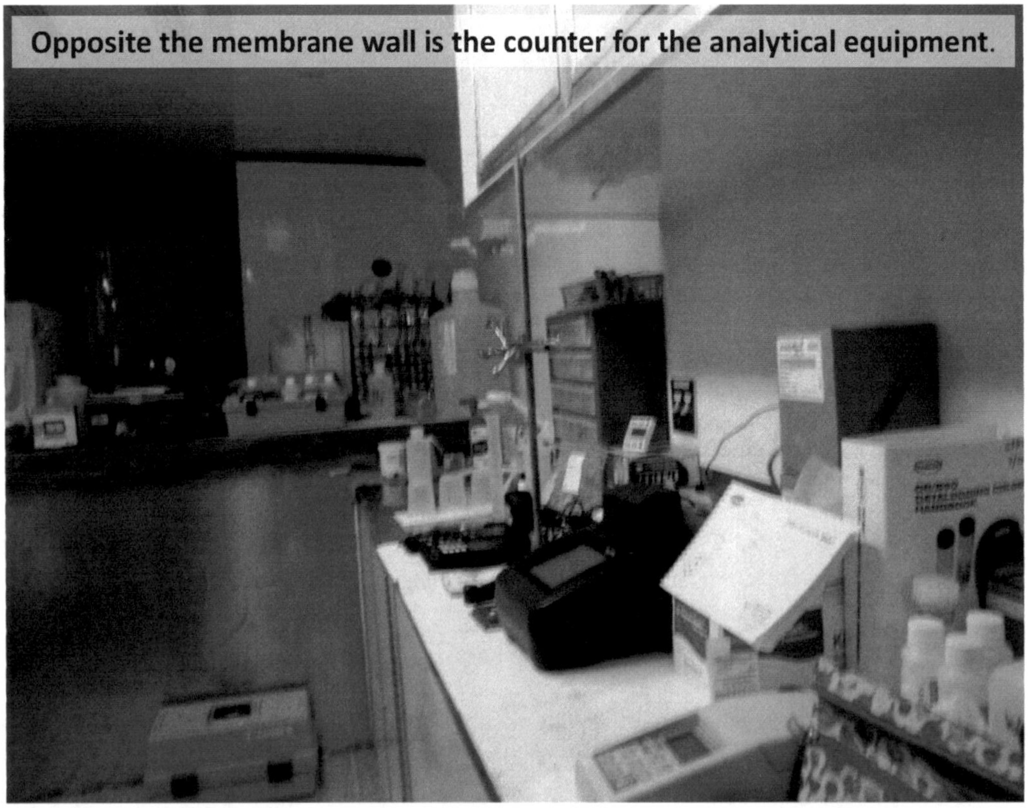

Opposite the membrane wall is the counter for the analytical equipment.

Field Frac Brine after Three Weeks

Treated **Un-Treated**

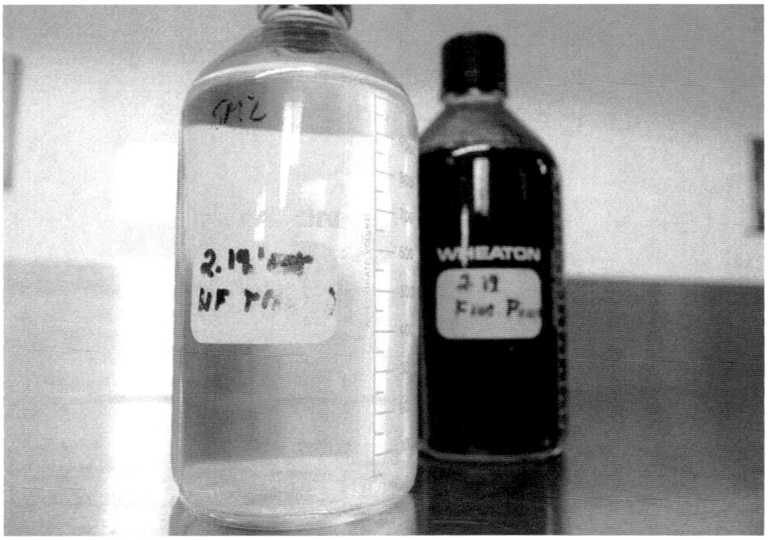

Analytes: Non Purgeable Organics

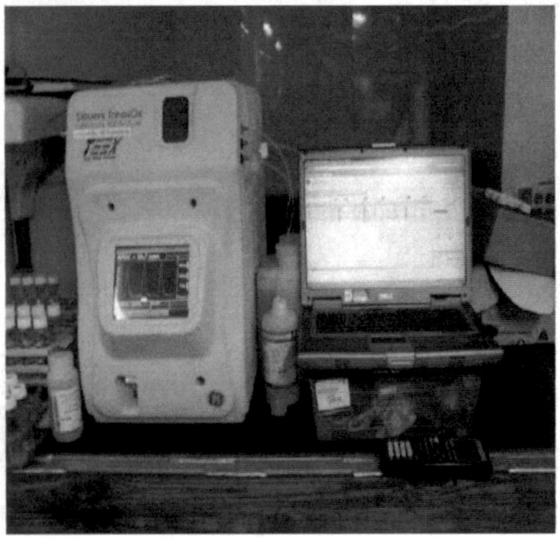

Figure 5. GE Analytical equipment for TOC analysis- optimized for hyper-saline brines

Figure 6 Samples Analysis in Mobile Laboratory

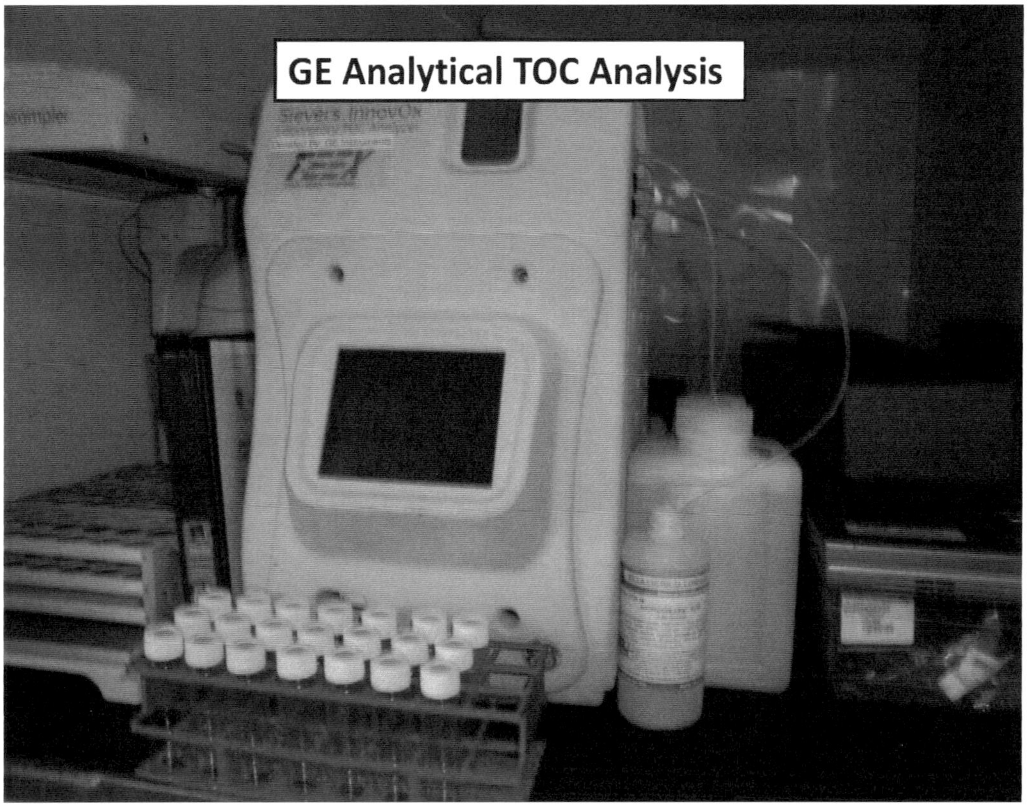

Hapsite GC/MS for BTEX/VOC

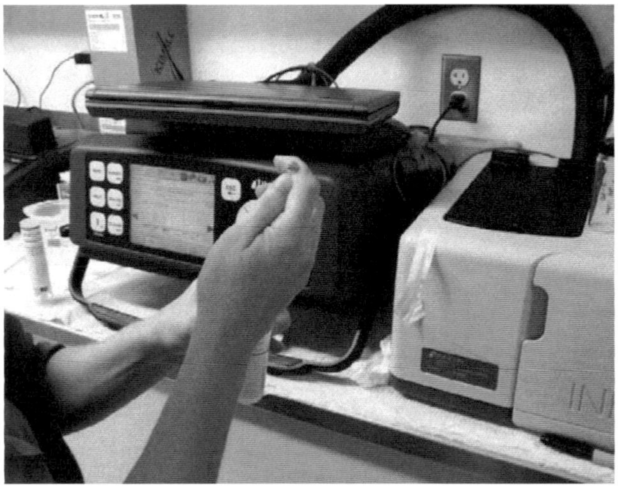

Questions?

Permian Basin, Unitah Basin, and Ohio Field Trials Open for Adoption

Harold Vance Department of
PETROLEUM ENGINEERING
TEXAS A&M UNIVERSITY

Sustainable Solutions for Flowback/Produced Water Challenges in Oil and Gas Fields

Chip Cagle
Veolia Water

VEOLIA ENVIRONNEMENT

The leader in sustainable environmental services

VEOLIA ENVIRONNEMENT

The only company in the world able to cover the entire range of environmental services

- $50 billion in revenue
- 312,000 employees in 74 countries
- 95,000 employees committed to water

VEOLIA WATER
The standard for water services
Revenue $18 billion

VEOLIA ENERGY
The standard for energy services and facility management
Revenue $10 billion

VEOLIA ENVIRONMENTAL SERVICES
The standard for waste management and resource recovery
Revenue $13 billion

VEOLIA TRANSPORTATION
The standard for managing safe and sustainable mobility solutions
Revenue $8 billion

Drilling & Fracking Process

- http://www.youtube.com/watch?v=73mv-Wl5cgg

- Marcellus Shale
  - Approx. 5,000,000 gallons of water used to frac each well
  - Approx. 1,000,000 gallons of water flows back over a 30 day period – (Flow Back Water)
  - After the well is put into production – a large portion of remaining 4,000,000 gallons is produced over the life of the well (Produced Water) – approx. 10 bbls (420gal)/day/well

Industry Challenges – Flow Back and Produced Water

- Increasing Transportation Cost
- Storage Cost – Frac Tanks, Impoundments, Tankage
- Flowback/Produced Water Processing
- Solid Waste Management
- Managing Total Cost
- Regulations Tightening
- Disposal Outlets are decreasing
- Best Available Technology

Goals: Reduce, Reuse, Recycle

Treatment Levels

- Level 1 – Minimal Treatment - **Reuse**
  - Total Suspended Solids Removal – TSS – Physical Chemical Treatment and or Filtration
- Level 2 – Specific Ion Removal + TSS Removal - **Reuse**
  - Target removal of ions like Iron, Calcium, Barium, Strontium, Metals – Physical Chemical Treatment
- Level 3 – Meet Discharge limits for **Disposal or Reuse**
  - Must remove Total Dissolved Solids (all ions)
  - Typical TDS levels in the Marcellus 120,000 to 300,000 ppm
  - Discharge limit <500 ppm
  - Typically Accomplished through application of thermal treatment when TDS levels >45,000ppm –(Evaporation/Crystallization Technology)
  - <45,000 ppm membrane technology is applicable

Client Decisions and Support

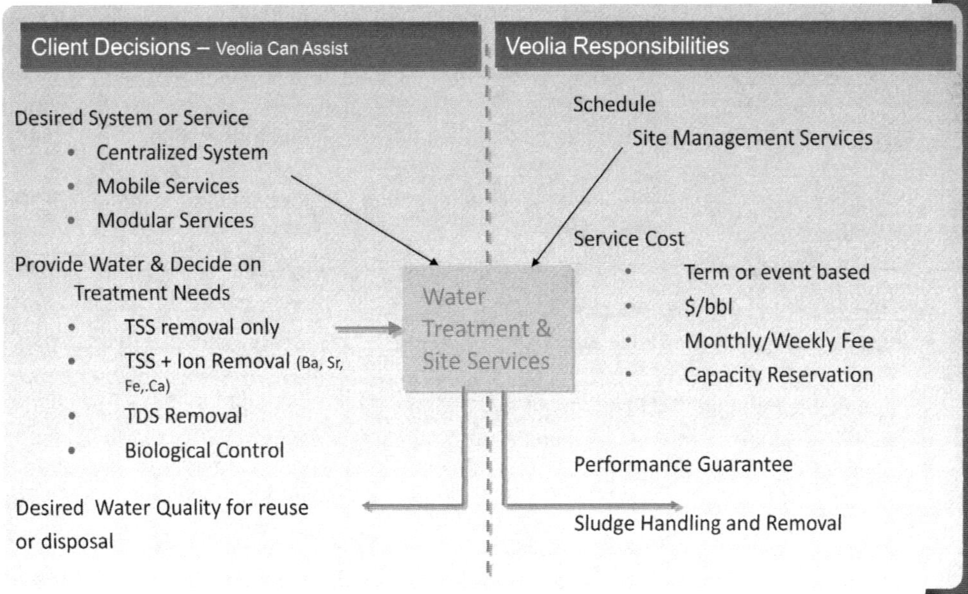

6

Shale Gas Plays, Lower 48 States

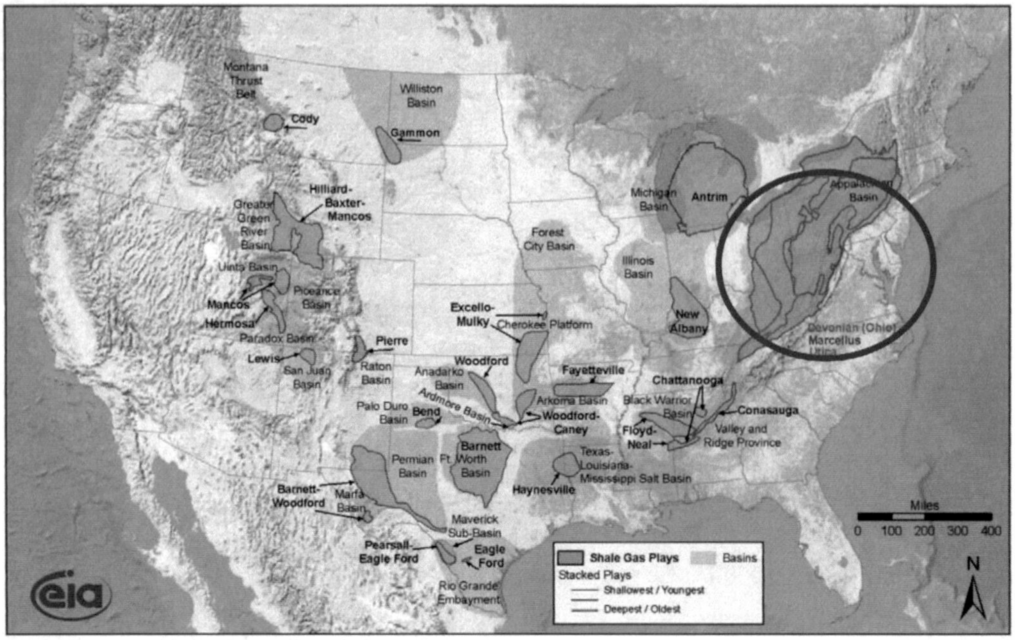

Source: Energy Information Administration based on data from various published studies
Updated: May 28, 2009

Total Water Management

- Includes
  - Collection
  - Transportation
  - Treatment
  - Mixing
  - Storage
- Reusing 100% of the flow back water and the majority of produced water

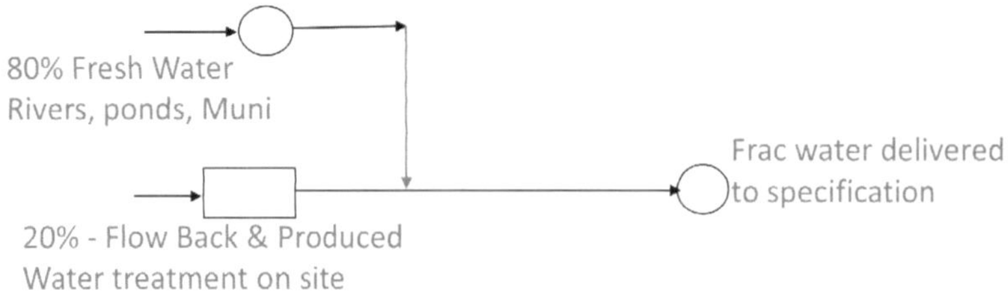

*Field Experience – **TSS and Iron Removal***

Case Study: Service Based

- Treatment in Marcellus with a major exploration company

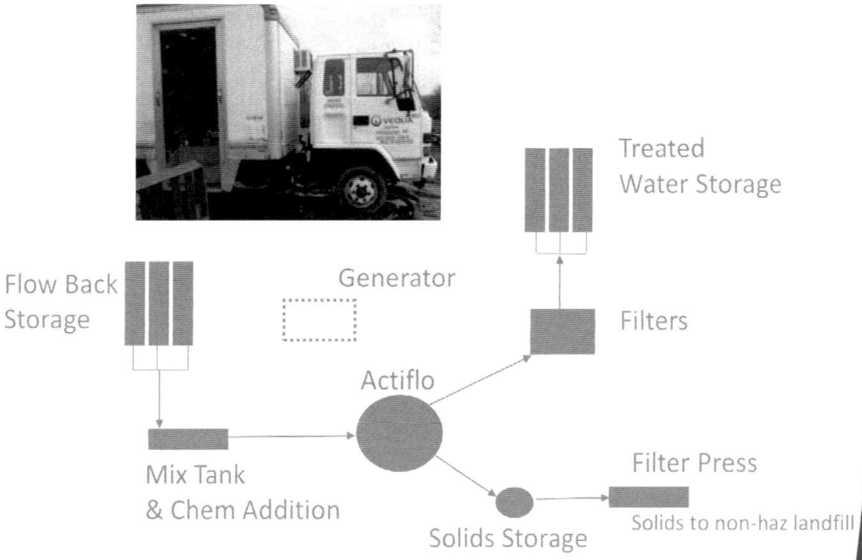

*Field Experience – **TSS and Iron Removal***

Case Study: Service Based

- Water and Solid Waste Handling
  - Water treatment to specific effluent quality
    - TSS <10 micron
    - Iron <10 ppm
    - Bacteria Control feeding biocide
  - Solids disposal to Veolia-owned land fill
  - Solids pass Paint Filter and TCLP tests

*Field Experience – **TSS and Iron Removal***

Case Study: Current Treatment Results - Water

Influent Turbidity	Influent Iron	Effluent Turbidity	Effluent Iron
NTU	ppm	NTU	ppm
170	47	<2	2.11
180	50	<2	<0.01
180	58	<2	<0.01
208	54.5	<2	<0.01
172	48	<2	2.36
179	50	<2	3.4
179	47	<2	3.4
175	45	<2	3.37
>400	54	<2	3.6
>400	70	<2	1.62

Flow rate = 300 – 750 gpm
Based on 10 days of operation at a Marcellus Shale site

11

*Field Experience – **TSS and Iron Removal***

Case Study: Current Treatment Results - Solids

- Challenge – Eliminate Hazardous Waste
  - Pass the paint filter test
  - Pass Toxicity Characteristic Leaching Procedure - TCLP
- Significant Cost Savings (Non-Haz vs Haz)
- Process stabilizes solid waste – ability to dispose of solid waste in non-haz landfill

12

Solids Waste Data - TCLP Testing

Extractable Metals	Result mg/l	Limit mg/l
Arsenic	<0.1	5
Barium	<0.5	100
Cadmium	<0.01	1
Chromium	<0.02	5
Lead	<0.1	5
Mercury	<0.001	0.2
Selenium	<0.1	1
Silver	<0.02	5
Volatile Orgainic Compounds	Result mg/l	Limit mg/l
Benzene	<0.02	0.5
Carbon Tetrachloride	<0.02	0.5
Chlorobenzene	<0.02	100
Chloroform	<0.02	6
1,4-Dichlorobenzene	<0.02	7.5
1,2-Dichloroethane	<0.02	0.5
1,1-Dichloroethylene	<0.02	0.7
Methyl Ethyl Ketone	<0.5	200
Tetrachloroethylene	<0.5	0.7
Trichloroethylene	<0.02	0.5
Vinyl Chloride	<0.04	0.2

System Photos

Shale Gas Plays, Lower 48 States

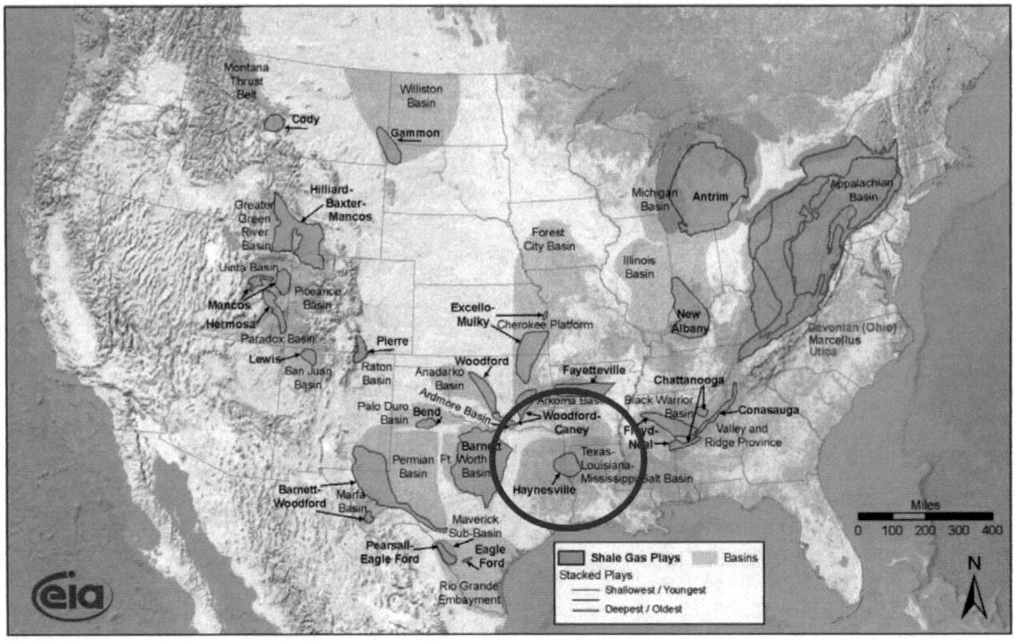

Source: Energy Information Administration based on data from various published studies
Updated: May 28, 2009

Field Experience – ion removal

Case Study: Mobile Water System for Frac Reuse

- Enables Flowback Reuse
  - Flow Rate up to 10,000 bpd
  - MULTIFLO™ Softening Technology
  - Associated Chemical Feed Skid
  - Fully Integrated System
- Removes Scale-Formers

Contaminant	Influent	Effluent
Calcium Hardness, ppm as $CaCO_3$	8,000 – 40,000	<50.0
Magnesium Hardness, ppm as $CaCO_3$	1,000 – 4,000	<50.0
Iron, ppm	25 – 200	<1.0
Manganese, ppm	5 – 10	<1.0
Barium, ppm	500 – 16,500	<10.0
Strontium	800 – 5,000	<10.0
Total Suspended Solids (TSS), ppm	100 – 1,000	<30.0

Oil Field - California

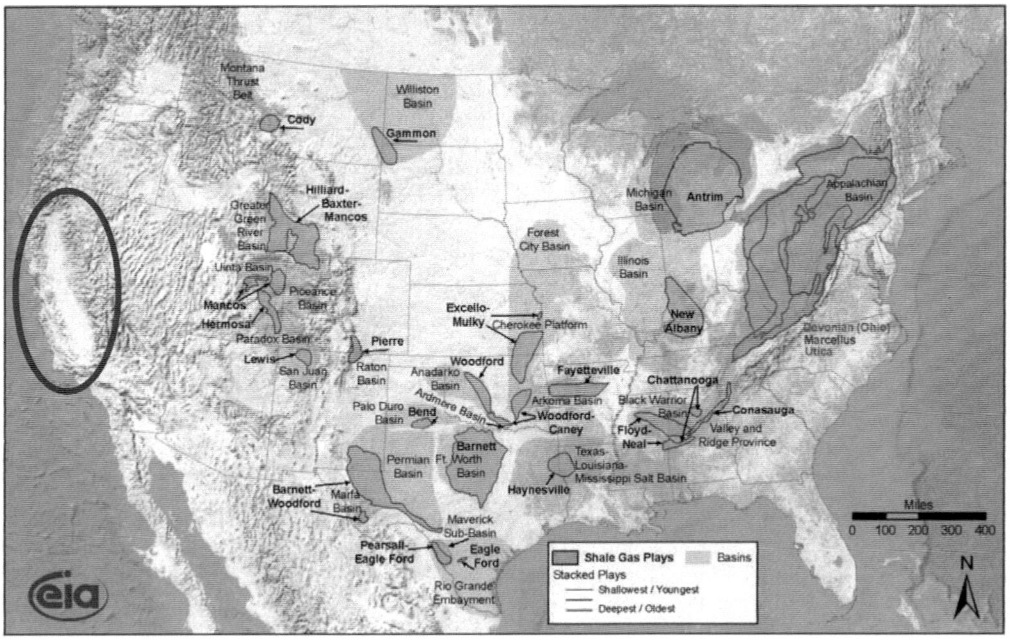

Source: Energy Information Administration based on data from various published studies
Updated: May 28, 2009

*Field Experience – **TDS Removal***

San Ardo Facility

Principle

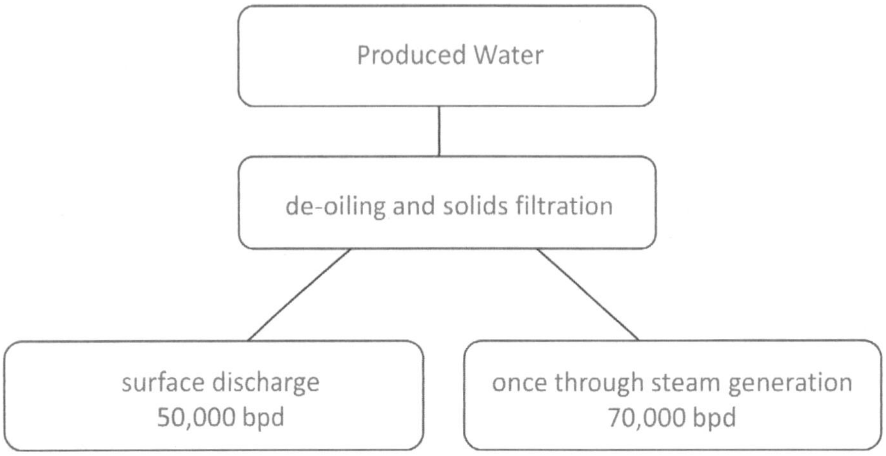

Treated Water Quality Required

- For surface discharge:
  - Total Dissolved Solids (TDS): <600ppm
  - Boron: <0.64ppm
  - Sodium: <100ppm
  - Chloride: <150ppm
  - Sulphate: <150ppm

- For steam generation:
  - Hardness: <2ppm as $CaCO_3$

Process Flow Diagram

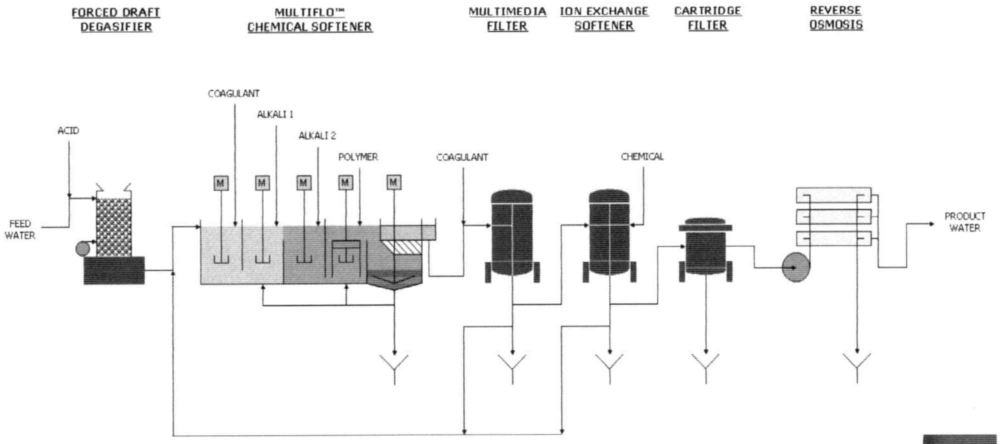

System Performance – Surface Discharge

Constituent	Produced Water	Discharge Specification	Final Treated Quality
TDS, ppm	7,000	600	120
SODIUM, ppm	2,300	100	43
CHLORIDE, ppm	3,400	150	11
SULPHATE, ppm	133	150	120
NITRATE, ppm	10.0	5.0	Non-Detectable
BORON, ppm	26.0	0.64	0.24
pH	7.5	6.5 ~ 8.4	7.0

System Summary

- A complex feed water source
- The worlds first membrane-based Produced Water desalination facility
- Successfully operated by Veolia Water for over 4 years
- RO performance data:
  - 80% recovery in first pass
  - 90% recovery in second pass
- CIP frequency:
  - First pass - once every 3 to 4 months
  - Second pass – non needed to date
- Sludge: 50% solids as a non-hazardous cake sent to landfill
- OPUS® is now a patented process

Water Reuse

*Field Experience – **TDS Removal***

San Luis Obispo, CA

- Similar to San Ardo – OPUS Technology
- 50,000 bpd
  - 20,000 bpd surface discharge
  - 25,000 bpd steam generator feed
- 90% recovery
- Influent Spec.
  - 1,950 ppm TDS,
  - 200 ppm total hardness
- Effluent Spec
  - Boron and Hardness <1 ppm
  - Silica <50ppm
- 10 yr term

Shale Gas Plays, Lower 48 States

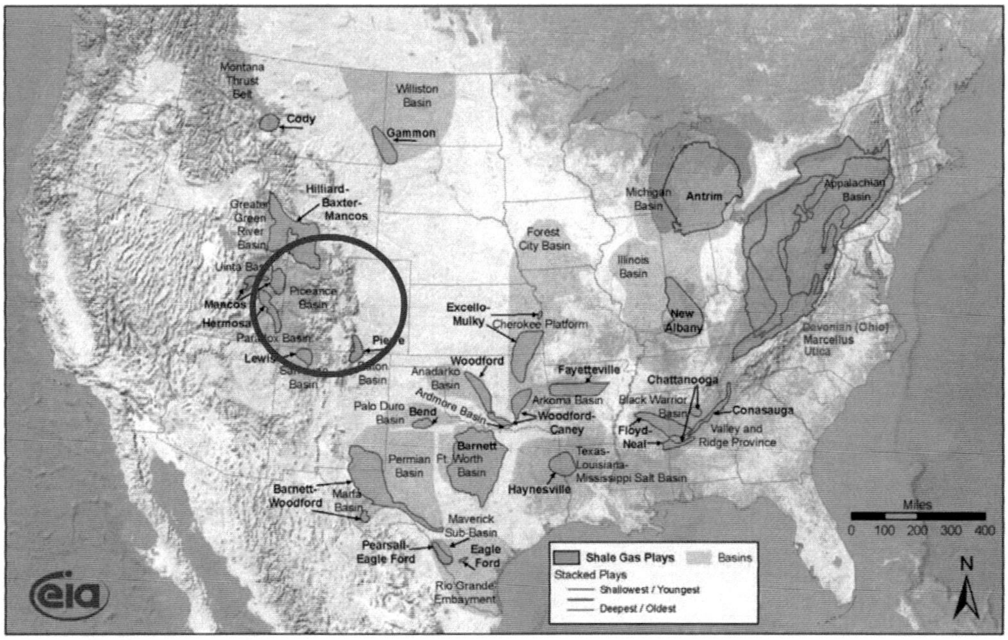

Source: Energy Information Administration based on data from various published studies
Updated: May 28, 2009

*Field Experience – **TDS Removal***

Piceance Basin – NW Colorado

- Modular System – Technology similar to San Luis Obispo
- Initial size 1-6000 bpd module with potential to expand to 4 modules or 24,000 bpd
- Influent spec
  - <15,000 ppm TDS
  - 750 ppm CBOD
- Effluent spec
  - <1270 ppm TDS
  - < 25ppm CBOD
  - < 1ppm Boron
- 75% recovery

VEOLIA
WATER

Thank you for your time!

Chip Cagle

864-350-2983

chip.cagle@veoliawaterna.com

AIChE Society for Water Solutions

- AIChE effort to help industry and industrial engineers more effectively exchange information and solutions to water management issues
  - Membership now an option for 2013 year (see AIChE dues notice!)
  - Corporate memberships also available
- Initial activities
  - Joint AIChE/AWMA workshop on water management in hydraulic fracturing Nov 1-2 during Annual Meeting in Pittsburgh
  - Series of workshops on Engineering Solutions to Water Grand Challenges under development
- For more information, contact
  - Danny Reible, Univ of Texas
    - reible@mail.utexas.edu
    - 512-471-0071
  - Ralph Exton
    - ralph.exton@ge.com
    - 215-942-3485
  - Darlene Schuster, AIChE
    - darls@aiche.org
    - 410-458-5870

Cockrell School of Engineering

Danny Reible
Bettie Margaret Smith Chair of Environmental Health Engineering
Director, Center for Research in Water Resources
University of Texas at Austin

WATER: CHALLENGING THE PRODUCTION OF GAS FROM SHALE

CRWR

Cockrell School of Engineering

Will Problems and Limitations of Water Limit Shale Gas' Potential?

(DATA FROM USEIA, 2010)

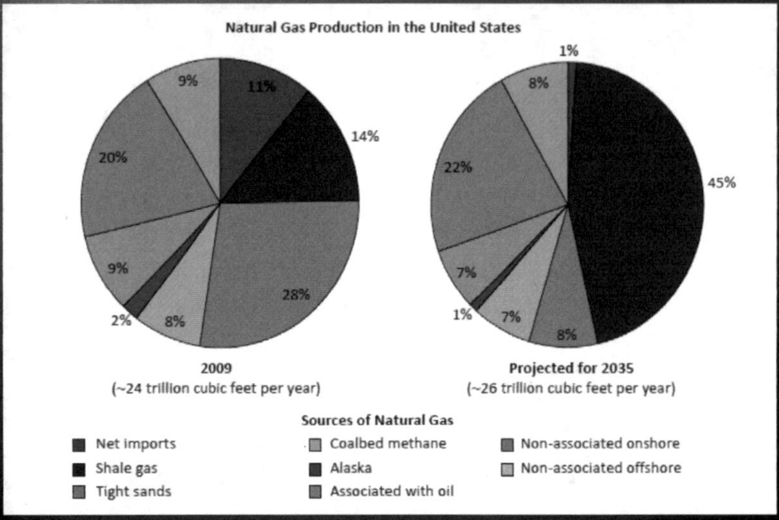

Cockrell School of Engineering

Public Perspective

- Strong desire to maintain rural residential character
- Strong desire to promote renewable fuels/reduce fossil fuels
- No mineral rights and concern for water
- No benefits/substantial potential cost from their perspective
- Result: No tolerance

Cockrell School of Engineering

Key Points from Day 1
Water Availability

- Overall demand for hydraulic fracturing may be small but timing and location of needed withdrawals may be significant
- Friction between individual ownership rights and community needs/desires
  - Water is a shared resource but ownership has historically not been shared
- Difficult balance between encouraging/allowing use and protection of water resources

CRWR

Cockrell School of Engineering

Key Points from Day 1
Water Management

- Avoid water sources that have other uses
- Change the fracturing fluid mix where possible instead of treating the water
- Recycle the flowback water
- Don't dispose of produced water on the surface
  - Loss to water cycle?
  - There are treatment options .... that are site specific..

CRWR

Cockrell School of Engineering

Key Points from Day 1
Water Treatment for Produced Water

- Techniques for oil/hydrocarbon removal
- Techniques for trace constituent removal (e.g. H_2S)
- Techniques for TDS reduction

- Often need to achieve all of the above
  - Expensive
  - Likely without specific drivers?
    - regulation, economics, necessity?

CRWR

Cockrell School of Engineering

What limits the "license to operate"?

- Water availability
  - Can we meet needs for water?
- Contamination of groundwaters
  - Can we produce gas without long-term negative impacts on a critical resource?
- Contamination of surface waters
  - Can we manage waters at the surface appropriately and dispose of wastes in a safe and efficient manner?
- *Managing the carbon footprint of gas production*
  - *..........save for another day.....*

CRWR

Cockrell School of Engineering

Specific Water Concerns of Hydrofracturing for Shale Gas

- Water availability
- Methane contamination of drinking water aquifers
- Spills and leaks during pre-fracture processing
- Casing failures during fracturing
- Fracturing allowing contaminant migration to drinking water aquifers
- Spills and leaks of flowback and produced water
- Inadequate treatment in POTWs
- Impacts on drinking water resources and utilities (e.g. increases in TDS, brominated DBP)

CRWR

Cockrell School of Engineering

Fracturing technology

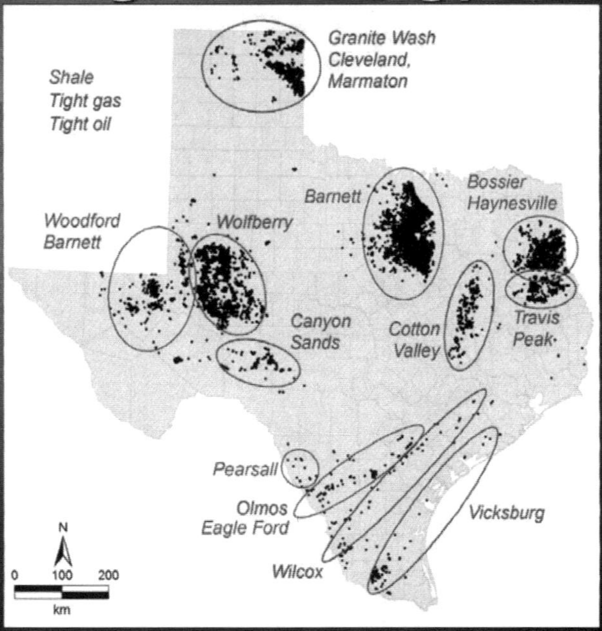

Nicot and Scanlon, 2012

CRWR

Cockrell School of Engineering

Water Demand- Median

- Barnett – 3 MM gal/well
- Haynesville – 6 MM gal/well
- Eagle Ford – 4 MM gal/well
- All approximately 1000 gallons/lateral ft of horizontal wells (80% between 500 and 1500 gal/ft)

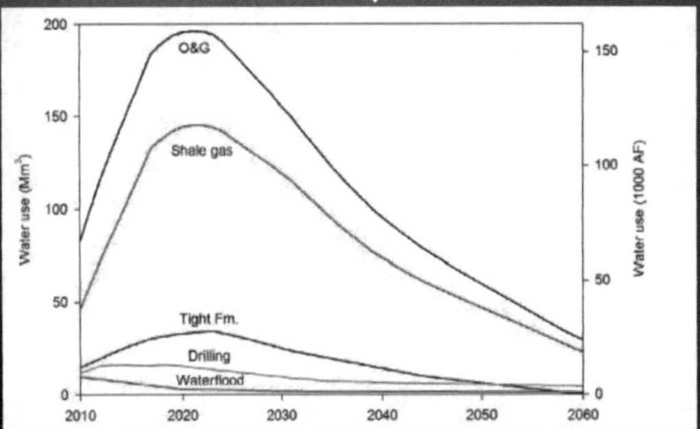

Nicot and Scanlan, 2012

CRWR

Cockrell School of Engineering

Water Availability – Eagle Ford

- Projected water demand – 5-6.7% of total water demand (Jester, 2011)
- Local dislocations possible (Nicot & Scanlon, 2012)
  - Projected HF as percentage of desired future pumping rates
    - Live Oak – 3.5%
    - De Witt – 8.3%
    - Karnes – 56.5%
    - La Salle – 66%
    - Dimmit – 130%
    - Webb - 136%

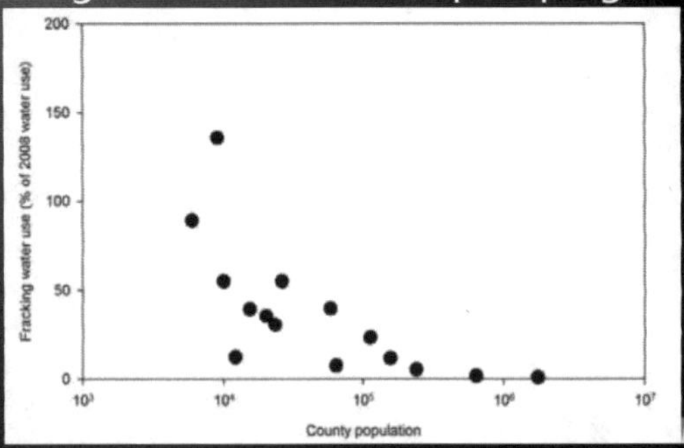

CRWR

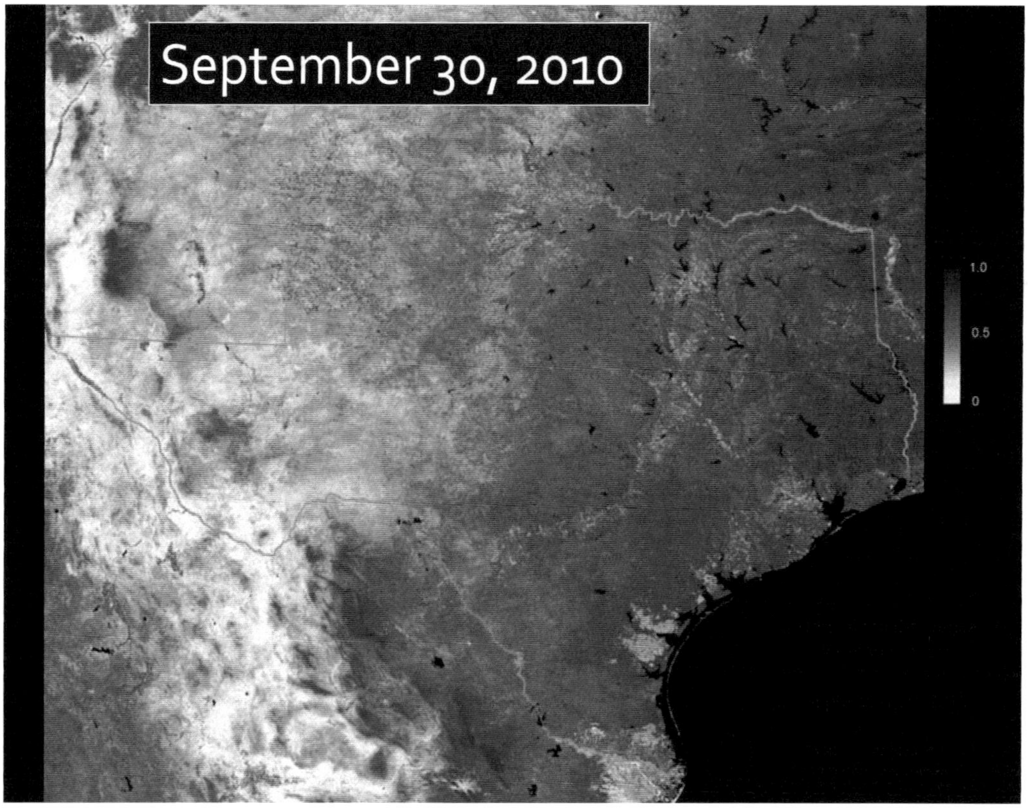
September 30, 2010
1.0
0.5
0

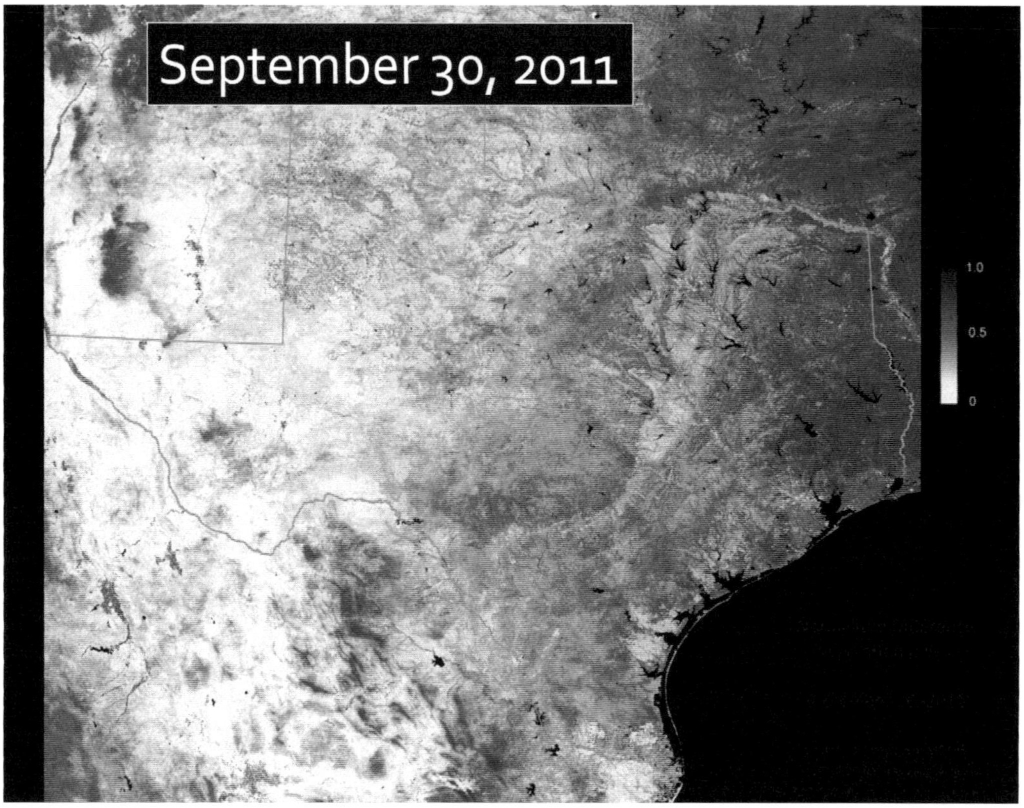
September 30, 2011
1.0
0.5
0

Drought Management is not the Solution
Texas Climate – Drought separated by occasional floods
How do we manage drought long term?
Supply – (Potential Demand – Conservation) + New Supply
Water
Gap
Natural Supply
Conservation
Water reuse
New Supply
Potential Demand
Cockrell School of Engineering

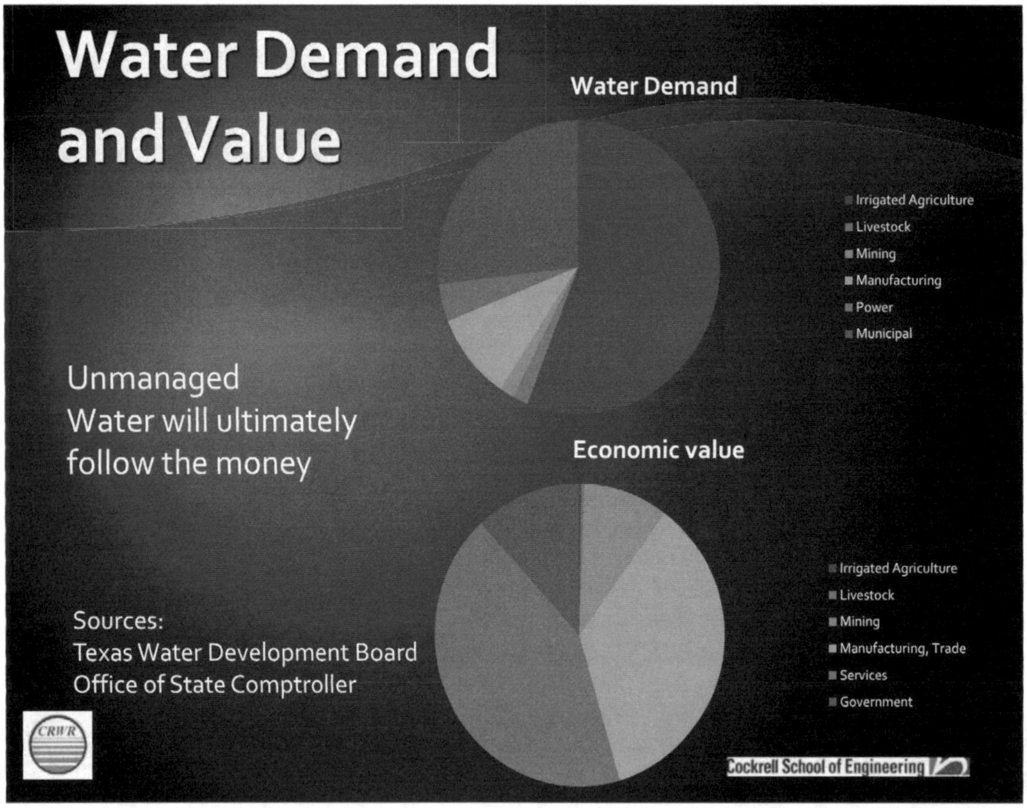
Water Demand
and Value
Water Demand
Irrigated Agriculture
Livestock
Mining
Manufacturing
Power
Municipal
Unmanaged
Water will ultimately
follow the money
Economic value
Irrigated Agriculture
Livestock
Mining
Manufacturing, Trade
Services
Government
Sources:
Texas Water Development Board
Office of State Comptroller
Cockrell School of Engineering

Conclusions- Water Availability

- Water will become increasingly monetized (and not just "valued")
- Flowback water in Texas will be increasingly recycled (following lead of Marcellus)
- Produced water will increasingly be treated to allow its retention in the surface water cycle

- What about environmental impacts on water?
  - Are near surface drinking water aquifers endangered?

CRWR

Cockrell School of Engineering

Well Containment Failure

- Frequency of blowout occurrence
  - Drilling: $6x10^{-5}$ per well drilled
  - During completion: $9.7x10^{-5}$
  - During wire-line logging: $6.5x10^{-6}$
  - Shallow gas exploration well
    - $6x10^{-4}$ for surface blowouts
    - $9.8x10^{-4}$ for underground blowouts
- Examples
  - Barnett Shale: 8-10 underground blowouts
  - Louisiana: transient pressurization of the shallow Wilcox aquifer
  - Ohio gas well: unexpected drilling into a shallow (~500m) high pressure natural gas pocket

CRWR

Cockrell School of Engineering

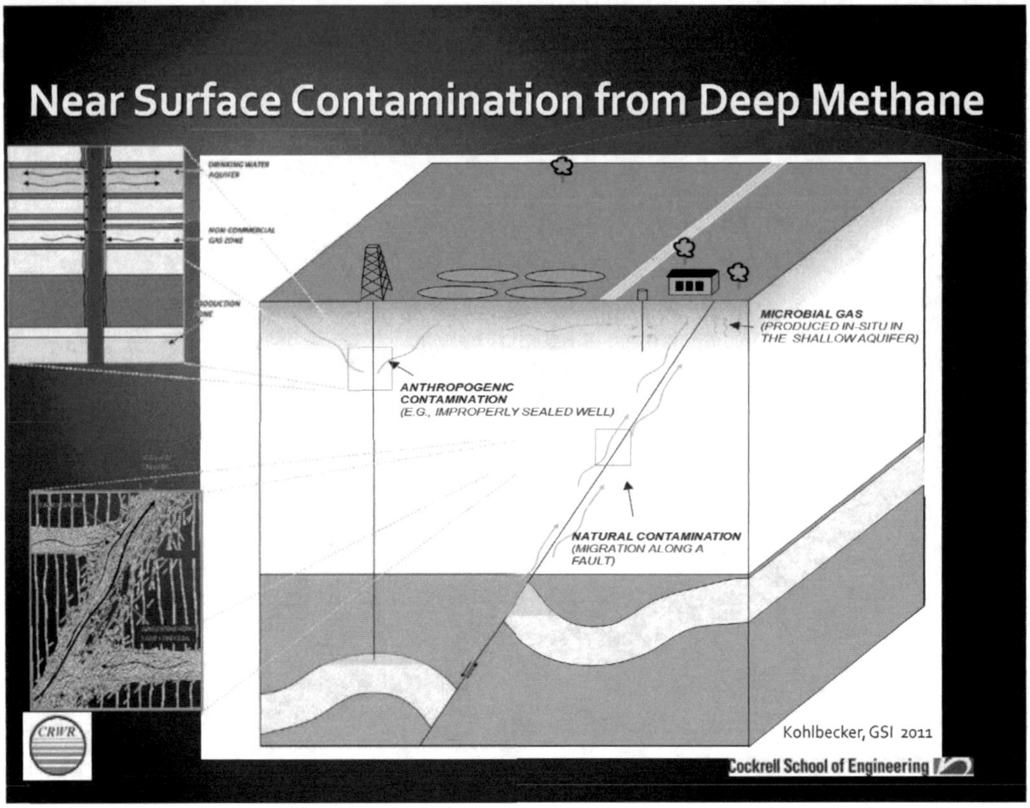
Near Surface Contamination from Deep Methane
MICROBIAL GAS
(PRODUCED IN-SITU IN THE SHALLOW AQUIFER)
ANTHROPOGENIC CONTAMINATION
(E.G., IMPROPERLY SEALED WELL)
NATURAL CONTAMINATION
(MIGRATION ALONG A FAULT)
CRWR
Kohlbecker, GSI 2011
Cockrell School of Engineering

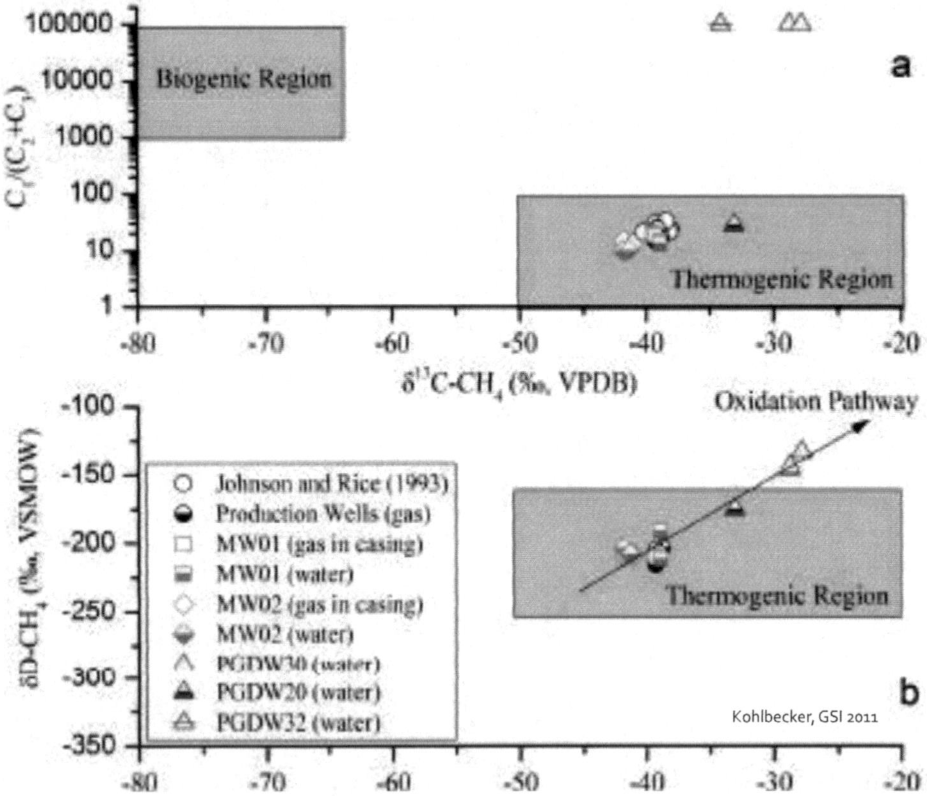
a
Biogenic Region
Thermogenic Region
C1/(C2+C3)
δ13C-CH4 (‰, VPDB)
b
Oxidation Pathway
Thermogenic Region
δD-CH4 (‰, VSMOW)
Johnson and Rice (1993)
Production Wells (gas)
MW01 (gas in casing)
MW01 (water)
MW02 (gas in casing)
MW02 (water)
PGDW30 (water)
PGDW20 (water)
PGDW32 (water)
Kohlbecker, GSI 2011

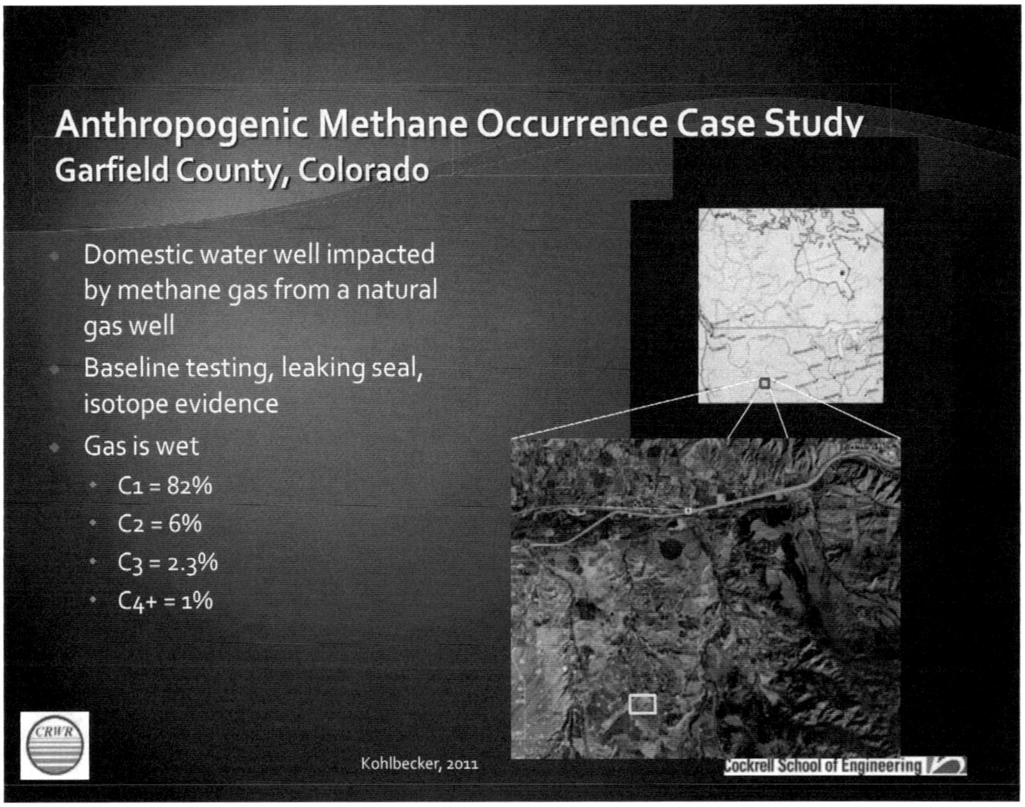
Anthropogenic Methane Occurrence Case Study
Garfield County, Colorado
Domestic water well impacted by methane gas from a natural gas well
Baseline testing, leaking seal, isotope evidence
Gas is wet
C1 = 82%
C2 = 6%
C3 = 2.3%
C4+ = 1%
CRWR
Kohlbecker, 2011
Cockrell School of Engineering

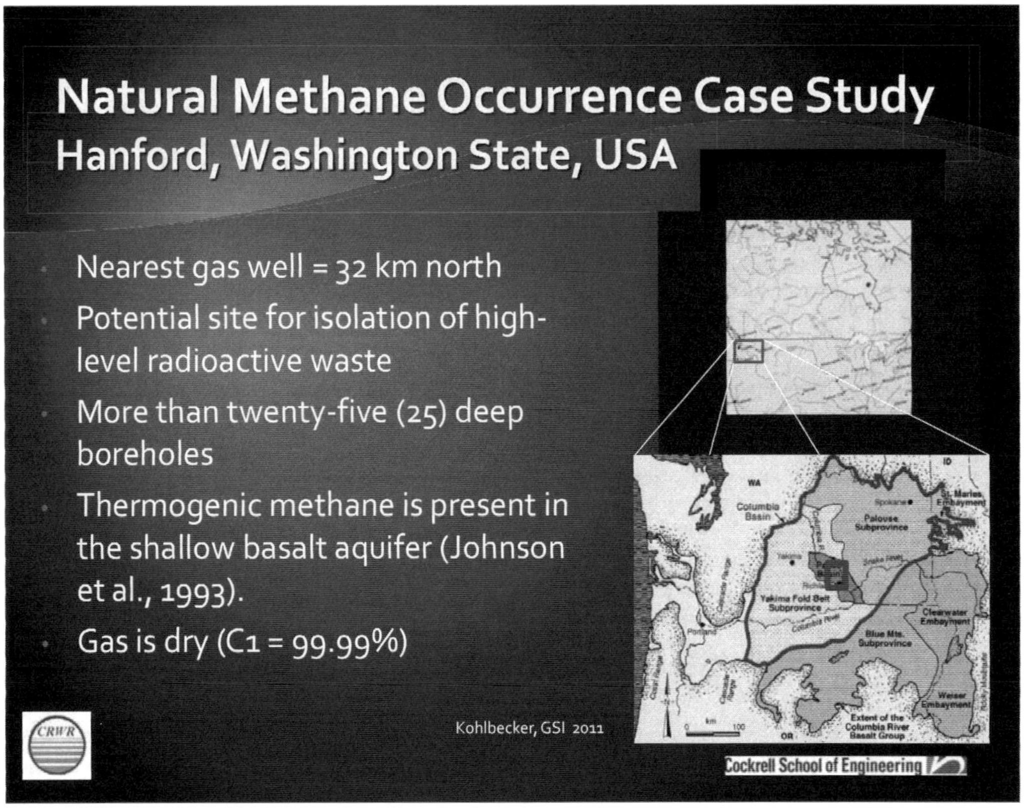
Natural Methane Occurrence Case Study
Hanford, Washington State, USA
Nearest gas well = 32 km north
Potential site for isolation of high-level radioactive waste
More than twenty-five (25) deep boreholes
Thermogenic methane is present in the shallow basalt aquifer (Johnson et al., 1993).
Gas is dry (C1 = 99.99%)
CRWR
Kohlbecker, GSI 2011
Cockrell School of Engineering

Bottom Line

- Hydraulic fracturing at depth may have minimal impacts on groundwater supplies although hydrocarbon migration is possible through natural faults
- Poor well construction or poor operations at the surface can impact groundwaters, surface waters and quality of life for nearby residents

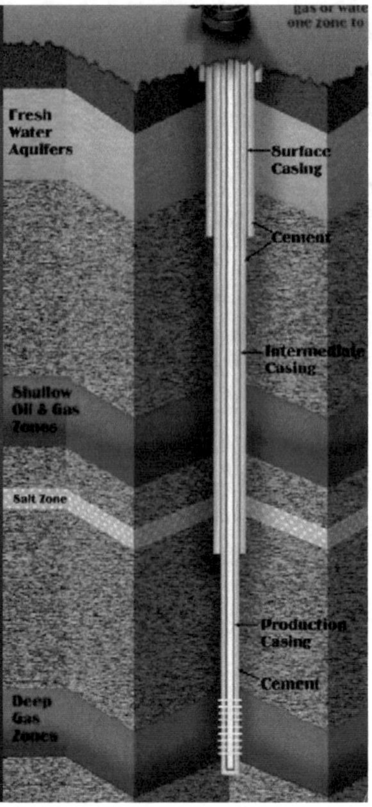

Perspective

- There is potential for contamination of drinking water supplies with any oil and gas activity
- Problems from drilling/production casing in shallow aquifers is likely more important concern than hydraulic fracturing at depth
- Surface management of injection and flowback/produced fluids and wastes can make or break perception of deep well hydraulic fracturing
- Increased drilling can mean more incidents/accidents even if rate decreases

Cockrell School of Engineering

Water Resource Management for Shale Energy Development

Fall 2012

Presented by David Yoxtheimer, P.G.

PENNSTATE

MARCELLUS CENTER
for OUTREACH & RESEARCH (M-COR)

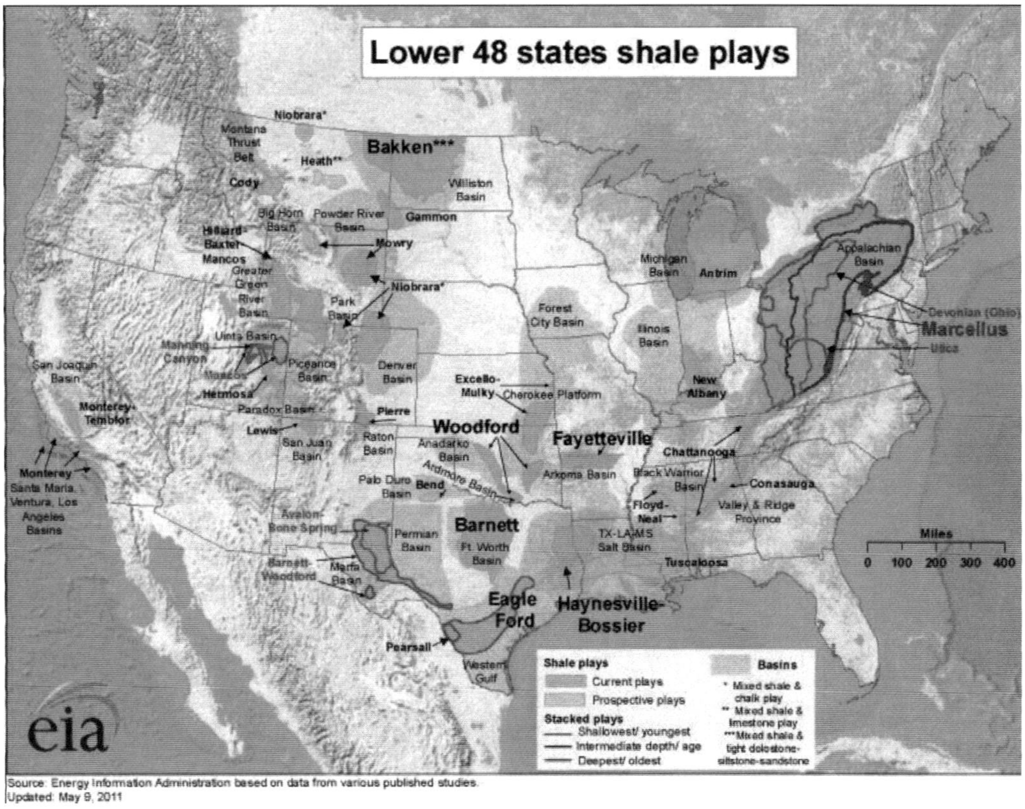

Source: Energy Information Administration based on data from various published studies.
Updated: May 9, 2011

Shale Energy: A Global Play

Figure 1. Map of 48 major shale gas basins in 32 countries

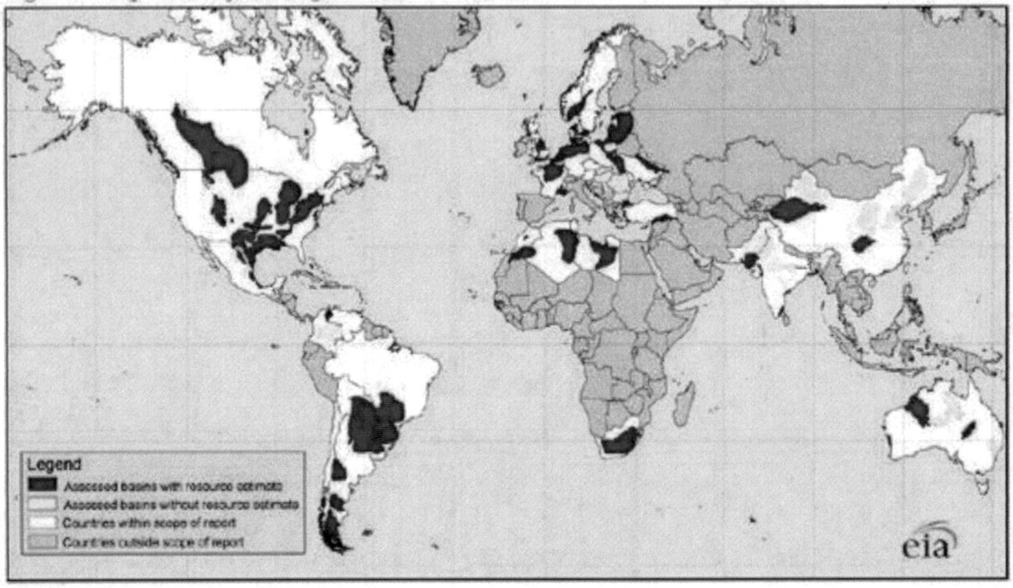

PENNSTATE Marcellus Center for Outreach and Research

Well Site in Operation

PENNSTATE Marcellus Center for Outreach and Research

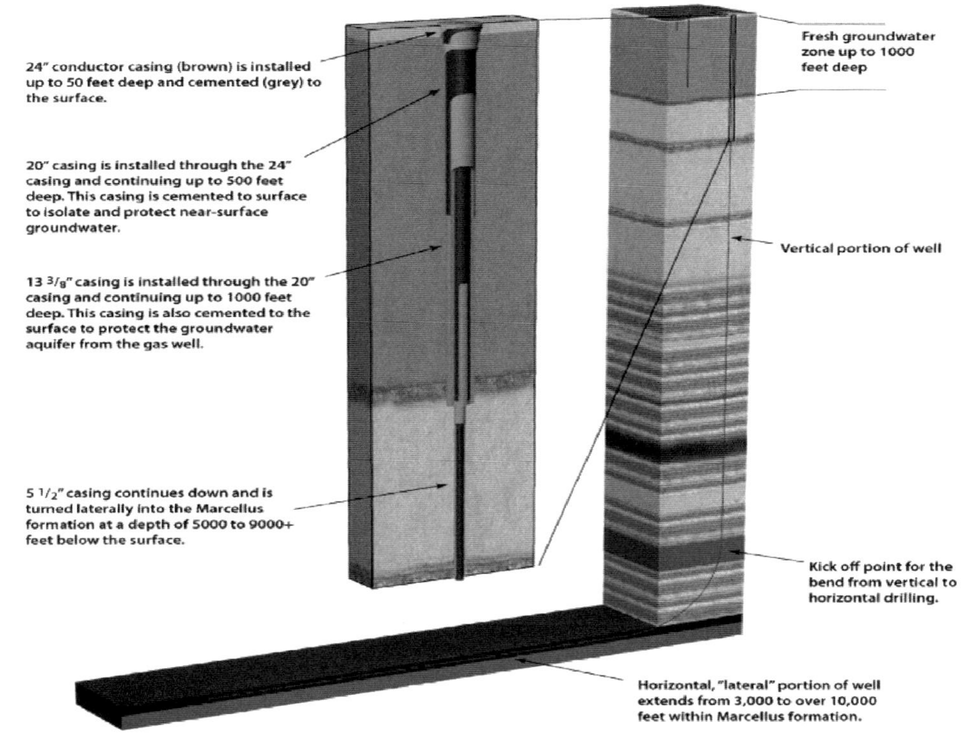

Marcellus Center for Outreach and Research, Penn State - www.marcellus.psu.edu

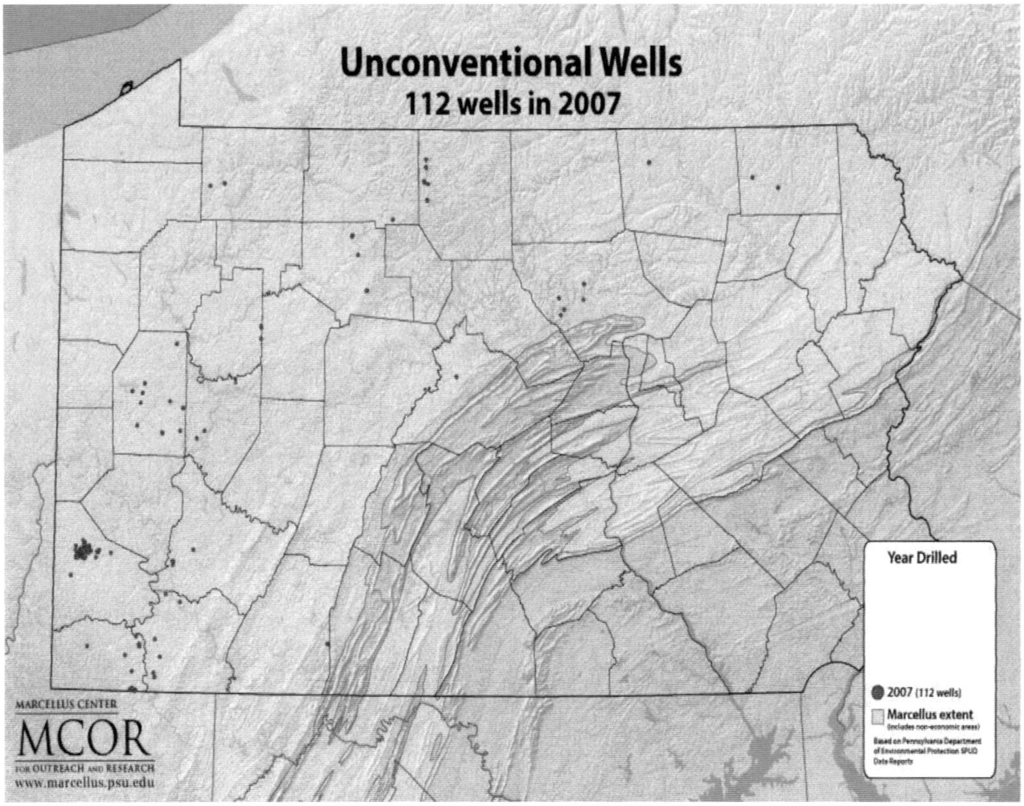

Shale Gas Development Water Use

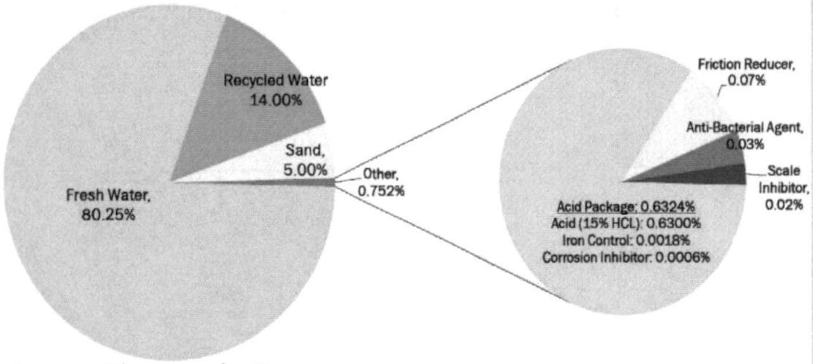

Source: Chesapeake Energy

- ~100,000 gall of water used during drilling
- 3 to 7 million gall of water used during hydraulic fracturing
- Approximately 8-10% of injected fluids return as flowback water

PENNSTATE Marcellus Center *for Outreach and Research*

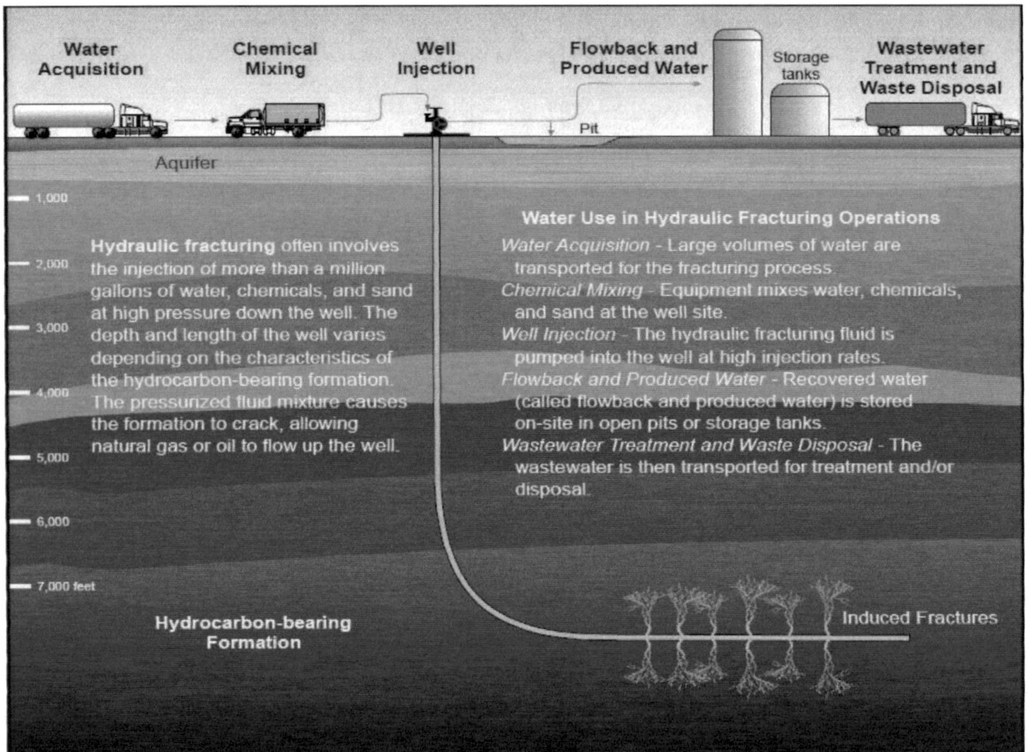

Source: Draft EPA Hydraulic Fracturing Study Workplan

PENNSTATE Marcellus Center *for Outreach and Research*

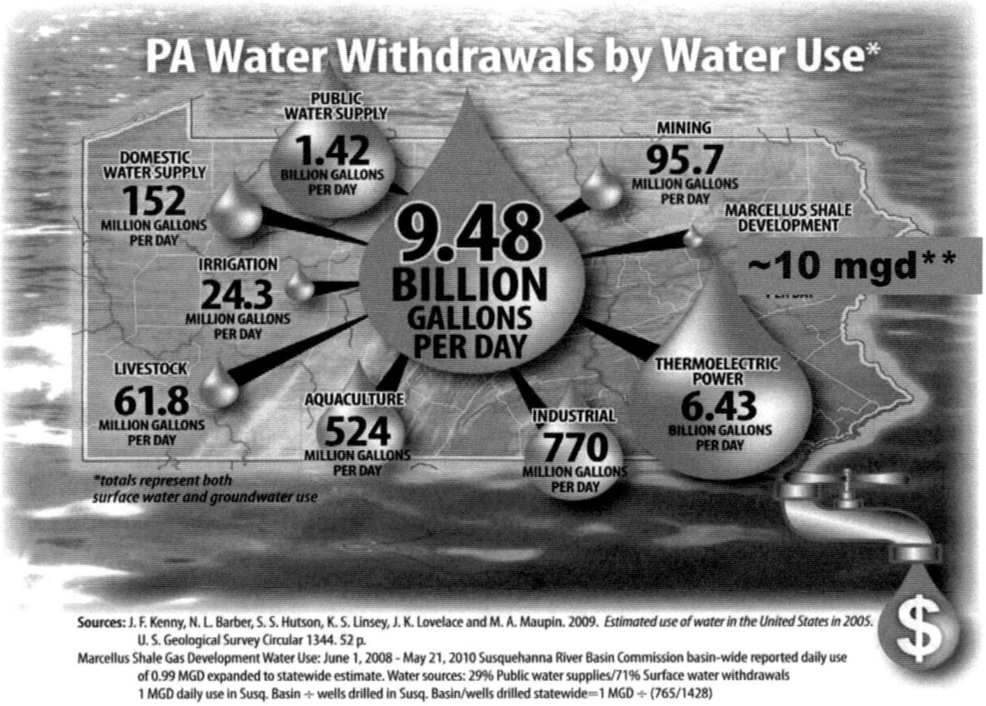

Source: Pa. Fish and Boat Commission **Estimated based on current SRBC/DEP data

Water Supply Considerations

Number of factors to consider:

- Access to water near the drilling project area
- Municipal vs. surface or groundwater supply
- Proximity to well site: piping vs. trucking
- Availability-seasonal or perennial
- Will pass-by flows be required?
- Water quality
- Drilling schedule vs. permitting schedule
- Budget

PENNSTATE Marcellus Center for Outreach and Research

Water Transport and Storage
MACK

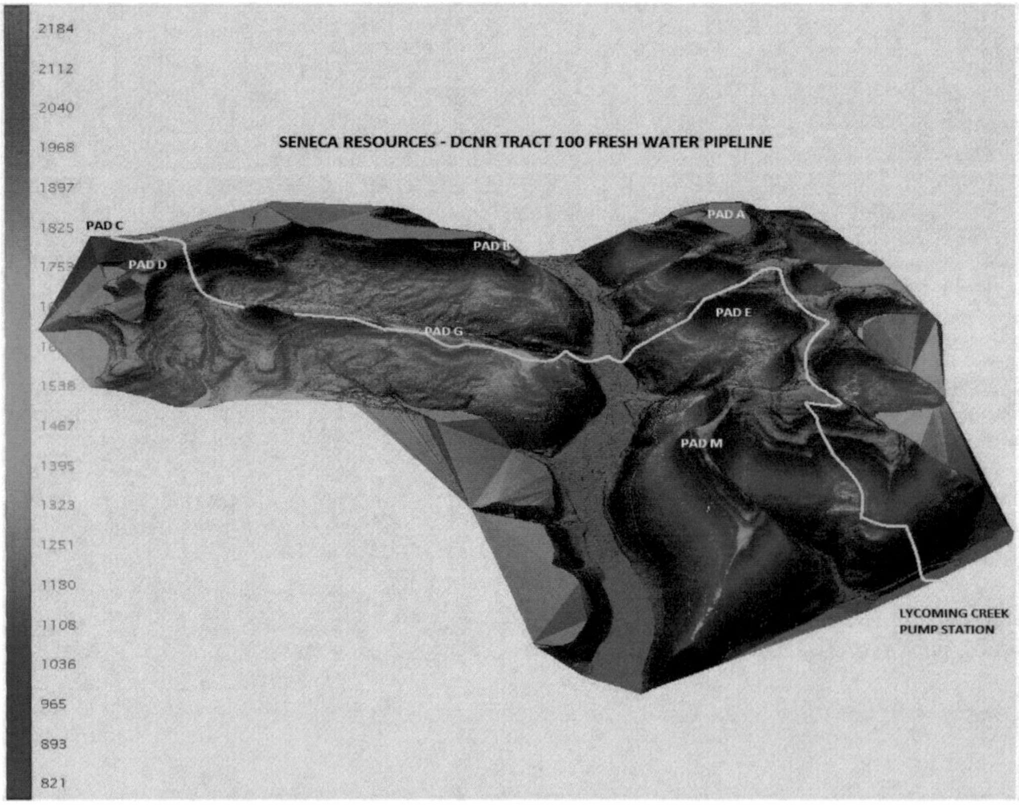
2184
2112
2040
1968
1897
1825
1538
1467
1395
1323
1251
1180
1108
1036
965
893
821
SENECA RESOURCES - DCNR TRACT 100 FRESH WATER PIPELINE
PAD C
PAD D
PAD B
PAD A
PAD E
PAD G
PAD M
LYCOMING CREEK
PUMP STATION

Intake Structure
SENECA RESOURCES
PENNSTATE
Marcellus Center
for Outreach and Research

Stream Flow Monitoring
PENNSTATE
Marcellus Center
for Outreach and Research

Intake Structure
PENNSTATE
Marcellus Center
for Outreach and Research

Temporary Impoundments
PENNSTATE
Marcellus Center
for Outreach and Research

Vertical Storage Tanks

PENNSTATE Marcellus Center for Outreach and Research

Closed Loop Drilling System

PENNSTATE Marcellus Center for Outreach and Research

Flowback Water Management Options

- Direct reuse (blending)
- On-site treatment w/reuse
- Off-site treatment w/reuse
- Off-site treatment and disposal (UIC wells)

~90% flowback and produced water reuse as of June 2012

Treatment standards for new or expanding treatment facilities

- TDS-500 mg/L
- Chlorides-250 mg/L
- Strontium-10 mg/L
- Barium-10 mg/L

In PA there are 7 plants that handled waste thru June 2012

PENNSTATE Marcellus Center for Outreach and Research

Flowback Water Quality Trends

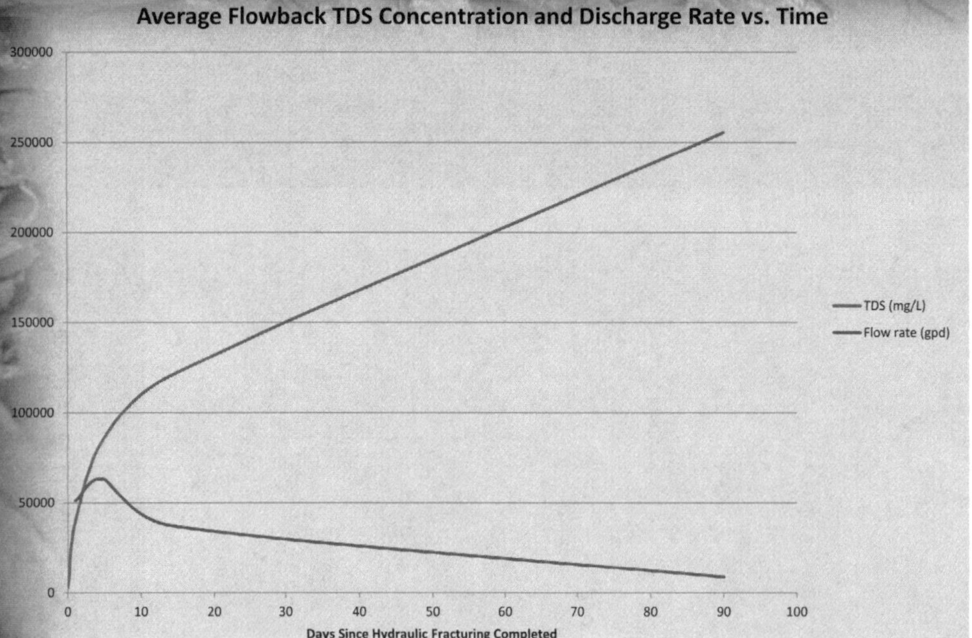

PENNSTATE Marcellus Center for Outreach and Research

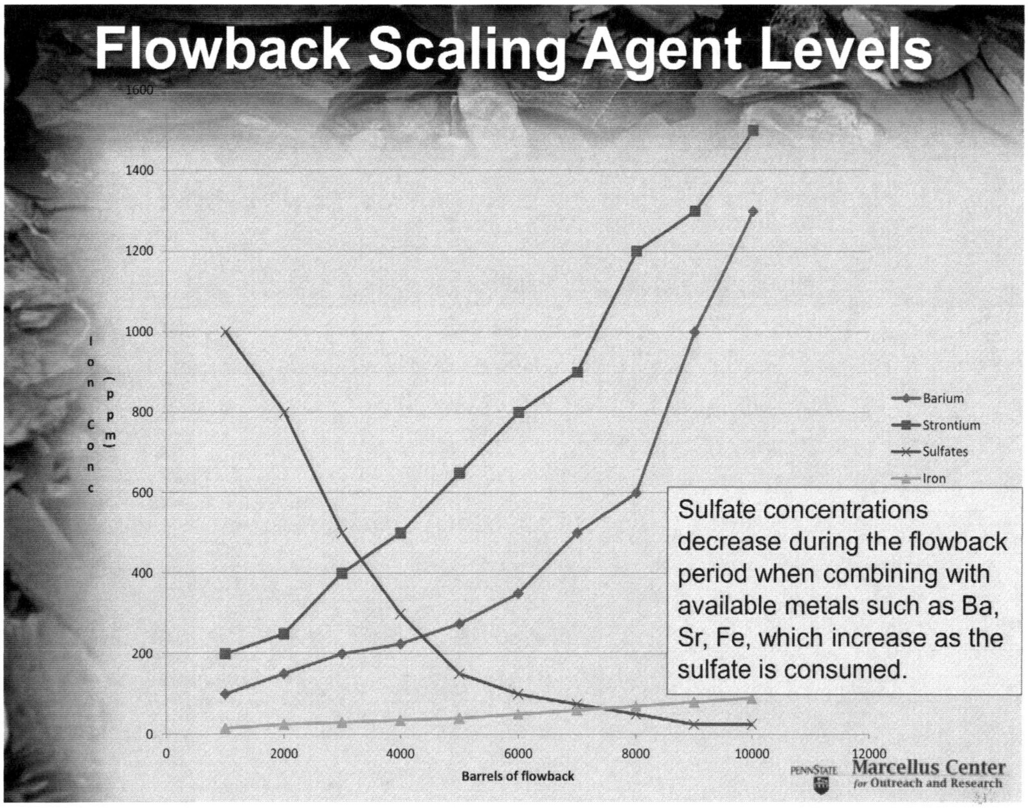
Flowback Scaling Agent Levels
Ion Conc (ppm)
Barrels of flowback
Barium
Strontium
Sulfates
Iron
Sulfate concentrations decrease during the flowback period when combining with available metals such as Ba, Sr, Fe, which increase as the sulfate is consumed.
Marcellus Center for Outreach and Research

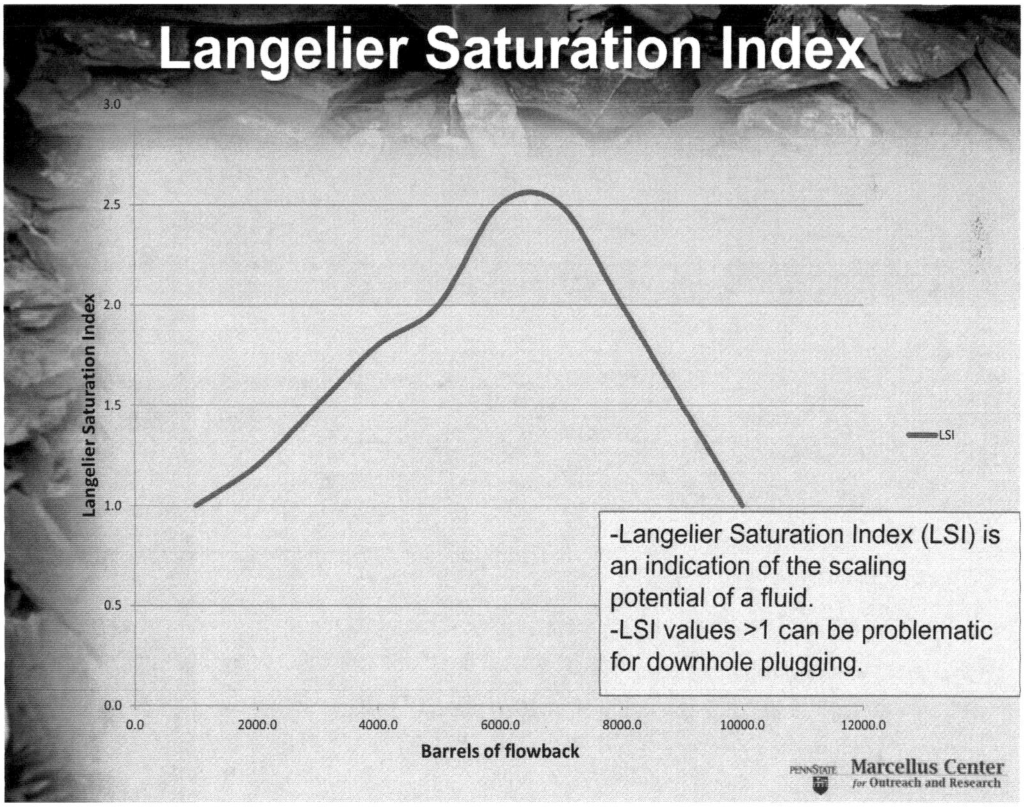
Langelier Saturation Index
Langelier Saturation Index
Barrels of flowback
LSI
-Langelier Saturation Index (LSI) is an indication of the scaling potential of a fluid.
-LSI values >1 can be problematic for downhole plugging.
Marcellus Center for Outreach and Research

Flowback Treatment Specifications

Example flowback treatment levels for recycling purposes per industry standards:

- Total cations 10 - 2,000 ppm range
  - Acceptable levels range from company to company
  - Primary focus on Ba and Sr, but Ca also a concern
  - Ba, Sr , Fe, Mn, Mg < 10 ppm
  - Ca <1,000 ppm
- Processed water sulfates levels <30 ppm
- TSS <30 ppm
- TDS is variable, >50,000 ppm can be acceptable

PENNSTATE Marcellus Center for Outreach and Research

Typical Treatment Scheme for Reuse

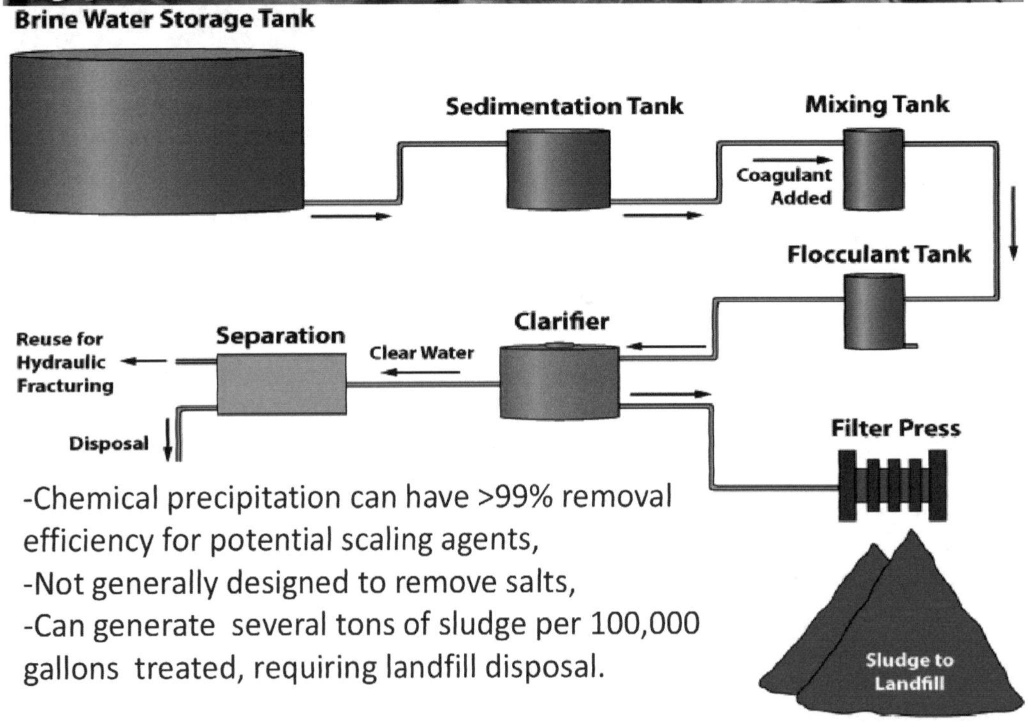

-Chemical precipitation can have >99% removal efficiency for potential scaling agents,
-Not generally designed to remove salts,
-Can generate several tons of sludge per 100,000 gallons treated, requiring landfill disposal.

High TDS Treatment Options

Technology	Maximum TDS (mg/L)
Reverse osmosis	45,000
Evaporation	100,000
Membrane distillation	250,000
Crystallizer	300,000

PENNSTATE Marcellus Center for Outreach and Research

Mobile Treatment Technology

Mobile treatment use increasing due to reduction in:

-Costs

-Trucking transport

-Fresh water use

-Environmental impact

PENNSTATE Marcellus Center for Outreach and Research

Flowback Storage

Flowback may be stored in lined impoundments for reuse
- Groundwater monitoring wells and leak detection
- Bird netting

Steel tanks are often used to store raw flowback or treated flowback
- Minimizes potential for spill

PENNSTATE Marcellus Center for Outreach and Research

Underground Injection Wells in PA

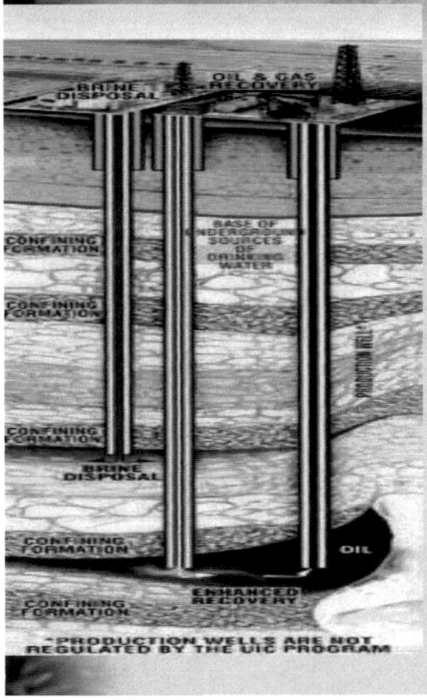

- Currently 5 operating injection wells in PA
- Relatively little use for Marcellus flowback disposal
- Difficult to find suitable target injection reservoirs in PA
- Induced seismicity is a concern
- EPA has primacy for UIC permit
- Two UICs recently approved in Warren Co.

PENNSTATE Marcellus Center for Outreach and Research

Water Management Trends

- **Use of alternative water sources**
  - Groundwater supply wells closer to drilling
  - Municipal wastewater use (3 plants in permitting process)
  - Acid mine drainage
- **Use of above ground temporary storage**
- **Flowback and produced water reuse is increasing**
  - ~90% industry-wide, was ~10% just a couple years ago.
- **Field recycling and fixed site treatment being utilized increasingly**
- **Closed loop drilling and fluids storage**
- **Lined well pads to minimize releases**
- **Piping water rather than trucking water**
- **UIC disposal well use increasing in OH**
  - Was ~20% in 2011, approximately 10% now
- **UIC well sites being pursued in PA**

PENNSTATE Marcellus Center for Outreach and Research

Thank you!

Questions?

PENNSTATE

MARCELLUS CENTER

for OUTREACH & RESEARCH (M-COR)

Water Treatment Technologies for Water Reuse in Hydraulic Fracturing

AIChE - A&WMA 2012 Workshop: Shale Oil & Gas E&P - Water Challenges & Opportunities, November 2, 2012

Aaron Johnson
Application Development Leader
The Dow Chemical Company

Dow.com

Dow Water and Process Solutions

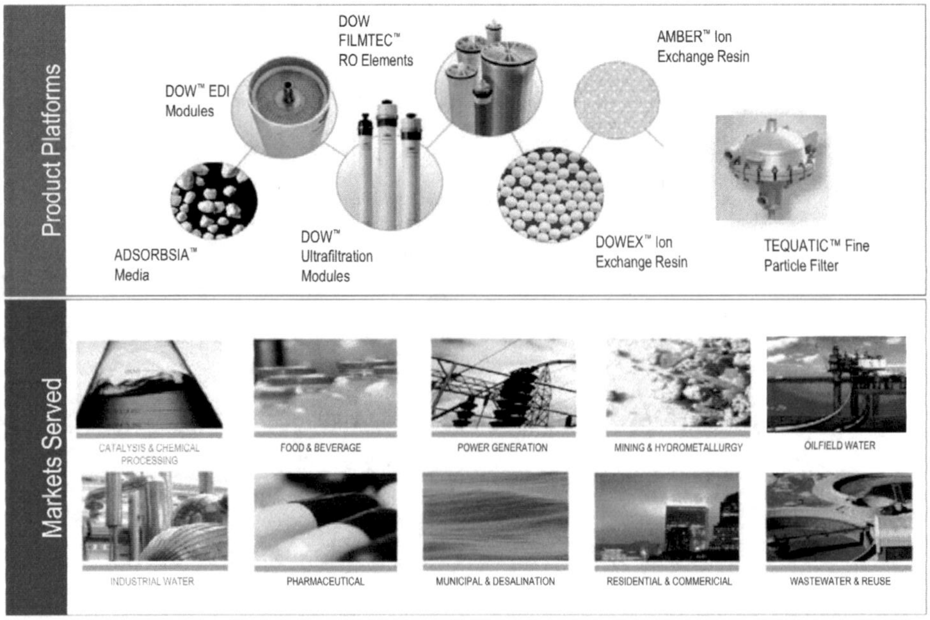

™Trademark of the Dow Chemical Company
Clean Filtration Technologies, LLC is a wholly Owned Subsidiary of The Dow Chemical Company

Oilfield "Produced Water"

Any water produced from crude oil and natural gas wells.

Volumes: The largest volume waste product in the hydrocarbons

Globally... around 275,000,000 bwpd (about 12,000 MGD)

By Field... typically 10,000 – 1,000,000 bwpd (0.4 – 40 MGD)

Quality: Highly variable difficult wastewaters

- Saline: TDS Ranges from "fresh water" to saturated brine... Two-thirds exceeding seawater
- Oily: Residual hydrocarbons vary by production (naphtha to tars)
- Complex & Dynamic : Equilibrium shifts, unsteady state flows, time dependant

Management Practices: Inextricably linked to their respective E&P Practices

- Enhanced Oil Recovery & Heavy Oil production with steam
- Offshore vs. Onshore considerations
- Unconventional Gas development and production

Today we are talking about water... in context of hydraulic fracturing... to produce gas and oil from Shales.

Shale Produced Water(s)

Shale Frac Flowback The portion of injected frac fluids that return to surface before production.
- Typically 10-30% returns quickly in 7-14 days with a rapid decline in quality & quantity

Shale Produced Water Typically refers to water produced from the Shale wells longer term.
- Significantly lower flow rates than flow back, and more consistent quality

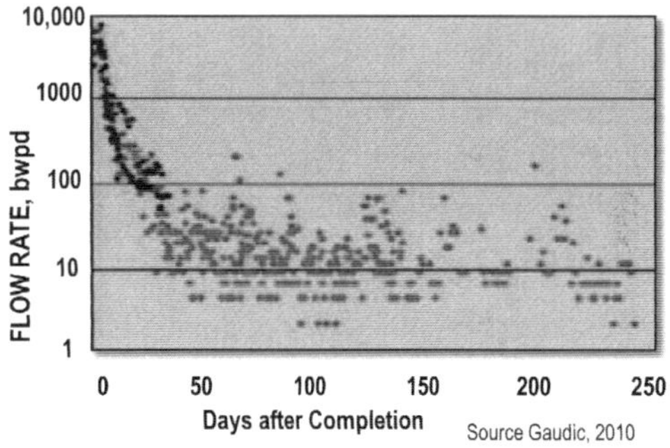

Source Gaudic, 2010

Shale Producer Water is characterized by:

- Formation solids, sand/proppants fines, drilling artifacts (cement, drilling muds, etc) in Flow Back
- High-MW polymers: Friction Reducers and/or Crosslinked "Gels"
- High salt content that increases over time
- Scale Potential of metals and divalent cations e.g. Fe, Ca, Mg, Ba, etc
- Residual light hydrocarbons: DRO, GRO, BTEX
- Naturally Occurring Radioactive Matter (NORM) e.g. Radium 226 / 228

Levels of Treatment Requirements for Shale Water

LEVEL 1 Reduce Suspended solids - Minimize impact on assets

Settle, Disinfect, Stabilize & Blend with fresh water
- Filter if necessary
- Monitor composition

Recycle or Deep Well Injection

LEVEL 2 Reduce Specific Ions - improve compatibility or performance

Requires Physical and Chemical Treatments
- Reduce Scaling potential of Ca, Mg, Ba, etc. or
- Meet specifications for fluid formulation (e.g. Boron, Fe and/or TDS)

Recycle / Reuse In The Oilfield

LEVEL 3 Purify Water - meet stringent quality standards

Advanced Treatment Systems -
- Removal of 99+ % salts, solids & organics to meet discharge quality
- Target specific trace contaminants and

Recycle, Beneficial Reuse or Discharge

Level 2 & Level 3 are driving needs more Advanced Tertiary Treatments.

Most Basic Treatment is the most common.

LEVEL 1 Reduce Suspended solids - Minimize impact on assets

Settle, Disinfect, Stabilize & Blend with fresh water
- Filter if necessary
- Monitor composition

Recycle or Deep Well Injection

Gravity Separation
- Majority of Solids settle in impoundment or storage Tank

Solids Hydrocyclones – Desanders/Desilters
- High Rate Removal of high concentrations of high s.g. solids
- Limited separation and recovery efficiencies for more neutrally buoyant Solids (lower s.g.)

Secondary Suspended Solids Removal
- Bag filters
- Depth filters

Automatic Self Cleaning Filters
- High solid capacity
- Absolute Filtration
- Low consumables

Example Water Specifications for Frac Fluid Recycle

TDS: TDS (or Chloride) levels specs are typically relatively high
High salinity may impact polymer hydration, but some is necessary

Scale Formers: Calcium, Magnesium, and Sulfate impact Scale Potential ; Hardness (Ca, & Mg) may interact w/ fluid chemistries, limit blending

Component	Unit	Marcellus Blend	Marcellus A	Marcellus B	Eagle Ford A	Eagle Ford B	Eagle Ford D	Foam Fluid
pH		8	6.5-7.5	6 - 8	6 - 8	5 - 9	5 - 9	5 - 8
TDS	mg/L	none	50,000	none	none	none	none	8,000
Cl	mg/L	25,000	-	20,000	30,000	10,000	60,000	4,000
Hardness, as CaCO3	mg/L		2,500					
Ca	mg/L	4200	1000		1,000	120	1000	100
Mg	mg/L	500	200		100	1000	1000	50
SO4	mg/L	124	-		1,000	500	500	200
Carbonate	mg/L	-	-		600	60	60	
Bicarbonate	mg/L	-	-		600			300
Iron	mg/L	15	5	20	25			4
TSS	Mg/L	1,500	-	50	30			
Boron					10	5	20	
Soluble Organics		5	-	25				

Boron & Iron: Boron Levels arise from Borate Crosslinked Gels. Must be removed to control crosslinking in new fluids, meet discharge standards

- For internal se ony

Treatments for Metals, Scale Formers, TSS

Chemical Softening

- Established Technology.
- Footprint & mobility limitations; sludge generation
- Driving advancement in high rate clarifier technology

Electrocoagulation

- Rapid oxidation reaction w/low footprint, Insitu flocculant generation
- Modular w/ potentially broad utility: demulsification, floatation, disinfection

Ozone-based systems

- May be used as reactors w/floatation
- Residual Microbial Control

Microfiltration & Ultra Filtration may reduce or eliminate settling time

These technologies generally do not reduce salinity of the produced water

TDS Reduction Options with Membranes

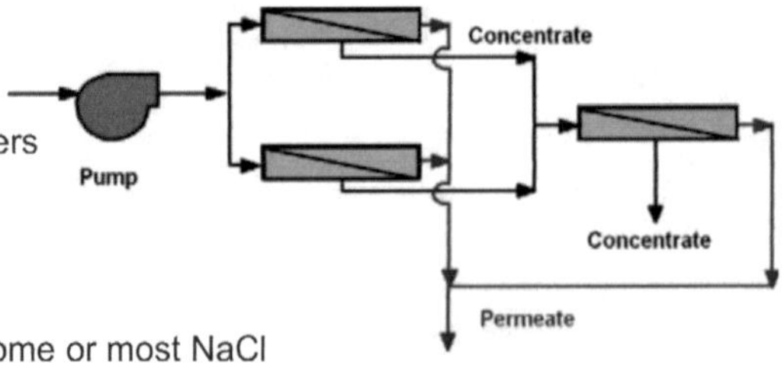

REVERSE OSMOSIS

- RO membranes reject 99+% of all ions
- Limited application for higher salinity Shale waters
- Water suitable for discharge applications

NANOFILTRATION

- NF membranes reject larger molecules, pass some or most NaCl
- Reject hardness from produced brines up to 120,000 ppm TDS with high Recovery
- More broadly applicable for frac water reuse applications w/saline source waters

Zero Liquid Discharge Systems May use Combinations of HIGH RECOVERY RO, NF & Thermal Evaporators and Crystallizers

Salt Concentration	Reverse Osmosis (RO)	Nanofiltration (NF)	Thermal Evaporation	Crystallization
Up to 20,000 ppm TDS	L3	L2 or L3	RO/NF Concentrate	L3 ZLD
20,000 to 45,000 ppm TDS	L3	L2 or L3	RO/NF Concentrate	L3 ZLD
45,000 - 120,000 ppm TDS	NF Permeate	L2 or L3	L3	L3 ZLD
Greater than 120,000 ppm TDS	n/a	n/a	L3	ZLD

Example of High Recovery RO w/Evaporation

A Desalination Facility treating Produced Water from Unconventional Gas Production

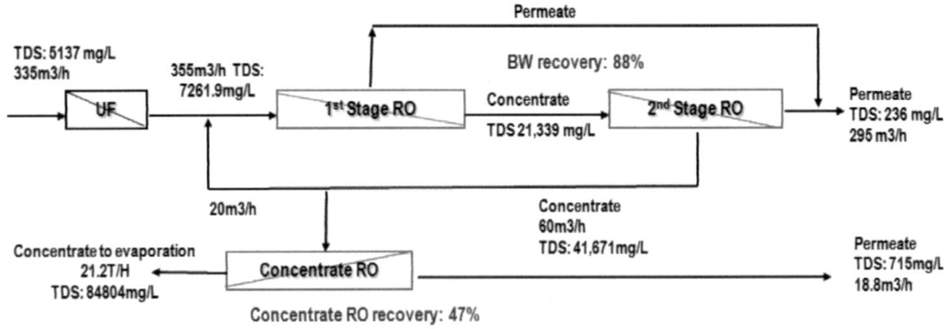

Whole system recovery: 93.6%

1st Stage: FILMTEC™ BW30HR-440i

2nd Stage: FILMTEC™ LE30-440i

Concentrate RO: SW30ULE-440i

Strict pre-treatment requirements are required for polymeric membranes

™Trademark of the Dow Chemical Company

Pretreatment for Membranes

Pretreatment is critical for sustainable operation of membrane systems

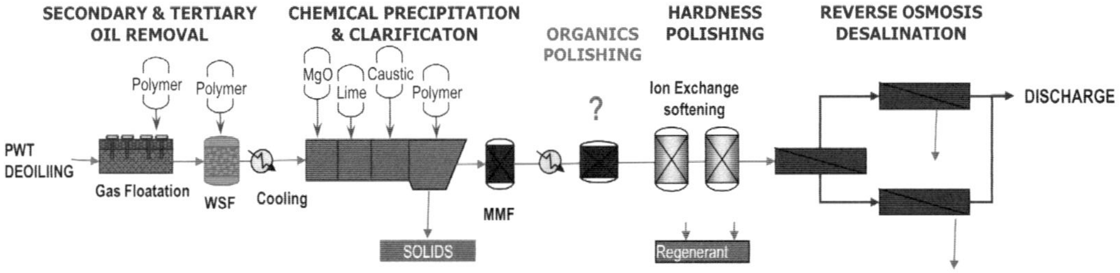

Produced Water Desalination PFD using conventional Pretreatment technology

Residual Oil and Organics are a Primary Concern

- **Natural organics** – primarily higher MW organic acids
- **Petroleum Hydrocarbons** - Oil & Grease, naphthenic acids, and water soluble components
- **Oilfield chemicals** - Surface active components: corrosion inhibitors, reverse emulsion breakers, solvents, etc.

Residual Oils Polishing

Technologies Demonstrated in Produced Water Membrane Desalination

TERTIARY OIL REMOVAL	ORGANICS POLISHING
Walnut Shell Filters (WSF)	Activated Carbon
Media Filters	Oil Adsorbent Filters
Modified Clays	DOWEX OPTIPORE™ Steam Regenerable Media*
Ceramic Membranes*	Modified Glass Media*

Water Soluble Organics are a critical concern

- **Gasoline Range Organics (GRO) and Diesel Range Organics (DRO), and BTEX**
- **Commonly imparted by Condensate and NGL production**
- **Foul, Swell and Degrade polymeric membranes**

Produced water samples from an OPTIPORE removal & recovery system. *Influent (left) effluent (middle) & recovered condensate (right)*

™Trademark of the Dow Chemical Company

Final Thoughts

- Development of Shale resources and other water-intensive E&P practices are driving adoption of water treatment technology for oilfield-produced water.
- Capabilities of available technologies are not the primary barrier to adoption in Shale Development. Expertise is needed to repackage for the oilfield and Implement complex technologies and treatments systems.
- Shale development will drive new applications & technology for Water.

Thank You

Aaron M Johnson - Application Development Leader
Dow Water & Process Solutions R&D
(989) 636-5817
amjohnson@dow.com

Produced Water Solutions, Inc.

November 2, 2012

Examples of Successful Recycle/Reuse Produced Water/Flowback Water Projects

12AIChE
Annual Meeting, Pittsburgh, PA

Summary of Topics

- **Produced Water Solutions, Inc. – Who We Are**
- **Current Method of Waste Water Disposal – costly and inefficient**
- **Recycling and Reuse of Waste Water – Treatment methods**
- **Cost comparison Recycle versus Disposal**
- **Value Proposition and Benefits of Treatment/Recycling**
- **Questions?**

 2

History of PWS

❑ Produced Water Solutions, Inc. was formed in 2008 as a Colorado Corporation

❑ Strategically located in Weld County, Colorado in the Niobrara/DJ Basin

(20 acre site with 15,000 sq ft of manufacturing, laboratory and office space)

❑ PWS has developed and deployed Mobile Treatment Units for the treatment of produced and flowback water for the oil and gas industry including:

- MTU-DW – Produced Water to Drinking Water 2009
- MTU-V.1 – Produced Water to Frac Water 2010
- MTU-V.2 – Produced/Flowback Water to Frac Water 2011

Proprietary & Confidential 3

Produced Water Treatment/Reuse for Fracturing Reduces Costs & Environmental Impact

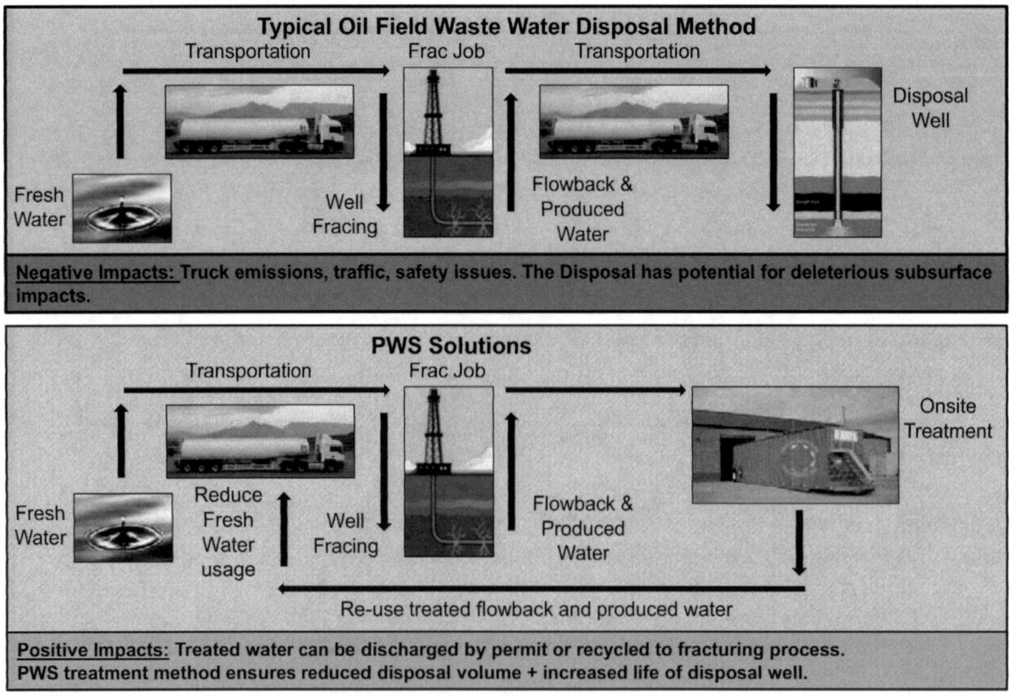

Average Produced Water Handling & Disposal Costs

- **Produced water handling and treatment represents an $18 billion cost to the Oil and Gas Industry in the U.S. alone.**
- **The cost of disposing of oil and gas produced water can cost as much as $13.00/barrel in some Basins.**
- **Trucking costs alone can be in excess of $10 per barrel (bbl)**

Source: "Modern Shale Gas Development in the United States," US Department of Energy, 2011

Mobile Treatment Options

**MTU-DW
Produced
Water to
Drinking Water**

**MTU-V.1
Produced/Flowback
Water to
Frac Water**

**MTU-V.2
Produced and
Flowback Water
to Frac Water**

MTU-V.3 for Produced and Flowback Water Treatment

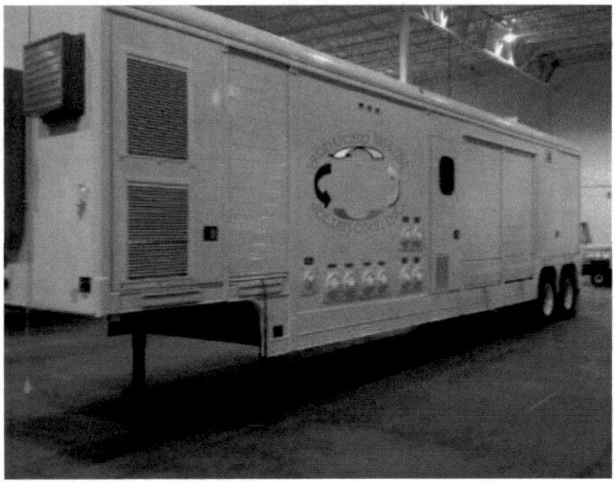

MTU-V.3 unit capable of treating 20,000+ barrels/day.

Results – Niobrara/DJ Basin Performance Produced Water to Frac Water

Produced Water → Treated Water

Marcellus
Disposal vs. Recycling Cost

All Costs are per bbl of water

Current Costs of Produced Water Disposal, Fresh Water Acquisition & Transport		vs.	Costs to Recycling Instead of Disposing	
Frac/Fresh Water Acquisition Costs	$0.25		Frac/Fresh Water Acquisition Costs	NA
Frac/Fresh Water Transport Costs	$2.00		Frac/Fresh Water Transport Costs	NA
Produced Water Disposal Fee	$2.00		Produced Water Disposal Fee	NA
Produced Water Disposal Transport Cost	$3.00		Produced Water Transport	$2.00
			Est. Treatment Costs	$3.00
Total Produced Frac Water Cycle Cost	*$8.25*		*Total Treatment Cost*	*$5.00*

Total savings per Barrel	$3.25
Percentage Savings	39%

* Cost savings do not show: reduced truck traffic, lower truck noise and other pollution, lower water usage, greater regulatory compliance, greater efficiency, lowered risk, etc.

Proprietary & Confidential 9

Summary - Recycling Treatment Features & Benefits

- Increased ROI. Treatment/Recycling is available at less cost than disposal and fresh water acquisition costs

- Mobile – We are where the needs are
- Modular – Multiple treatment processes are available to meet specific "fit for purpose" specifications
- Reduce transport costs and emissions (CO2, VOC, methane, etc.)
- Reduce/mitigate environmental impacts
- Relieve stress on community and oilfield infrastructure

Proprietary & Confidential 10

Produced Water Solutions Inc.

USA

"Why truck when we can treat for less?"

Produced Water Solutions Inc

2425 W. I-25 Frontage Road
Erie, Colorado 80516

Phone: (303) 215-9595
FAX: (877) 531-7044
Email: info@produced-water.net

Website: www.produced-water.net
Twitter: @Cleaner_Water
Facebook: Produced Water Solutions, Inc.

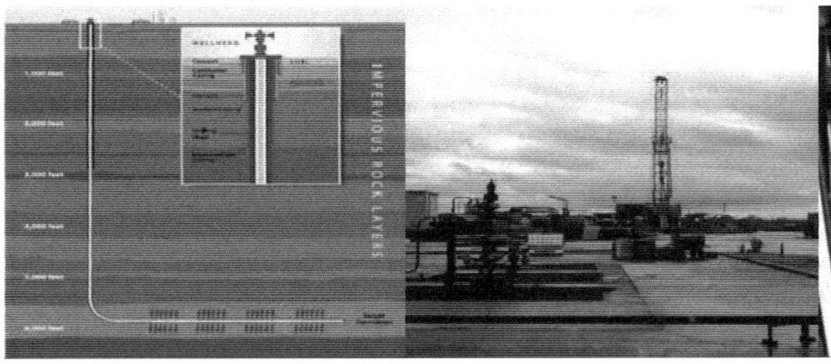

AIChE/AWMA 2012 Workshop: Shale Oil & Gas E&P - Water Challenges & Opportunities

Strategic Decision-Making for Water Management

Tekla Taylor, RG
US Energy Initiative Leader
Golder Associates Inc.

Golder Associates

Water Use for Shale Gas Development

- AVAILABILITY
  - Shale Gas Represents ~1% of Total Water Use
- COMPETITION
  - Current and future industrial, domestic, recreational, and habitat uses
  - Drought
- OWNERSHIP
  - Water is a Right

Source: USGS

2

Golder Associates

Approach to Water Management

- View Water Management as a Supply Chain or Lifecycle
- Identify local limiting factors
- The weakest link in the chain presents the greatest risk
- Assess risk
- Identify alternatives
- Use strategic decision-making to evaluate feasible, economical solutions
- Implementation of Water Management Strategy

3

Golder Associates

Why Develop a Water Management Strategy?

- Achieve consistency in water management and planning across all plays
- Reduce risks associated with securing water resources and handling wastewater
- Gain commitment from all levels of management and operations
- Demonstrate sustainabiity to stakeholders

4

Golder Associates

What Affects a Water Management Strategy?

- Operational Impacts/Inputs (what really works best)
- Innovation (e.g. treatment technologies or frac fluid chemistries)
- Stakeholder influences
- Regulations
- Regional and local economies

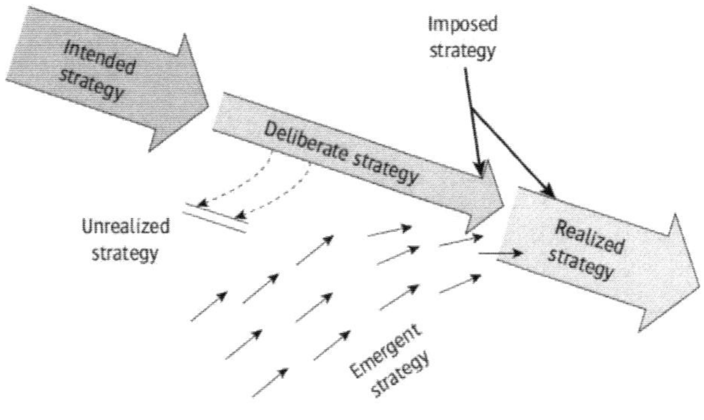

Source: Mintzberg and Waters, 1985

5

Golder Associates

Where Should Strategies Be Developed?

- Corporate Level
- Business Level
- Functional Level

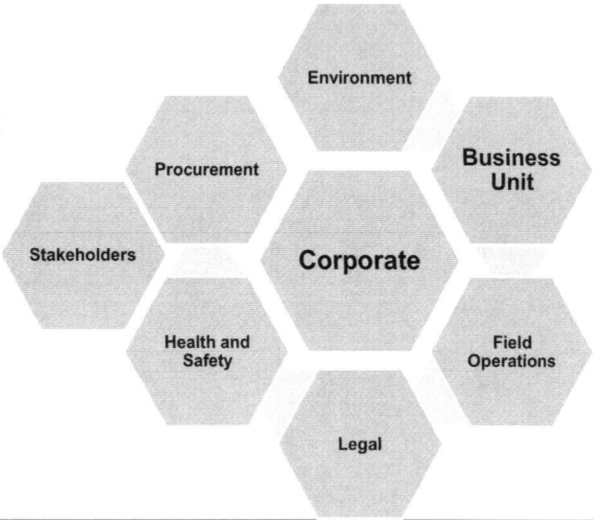

6

Golder Associates

The Need for Strategic Decision Making Tools

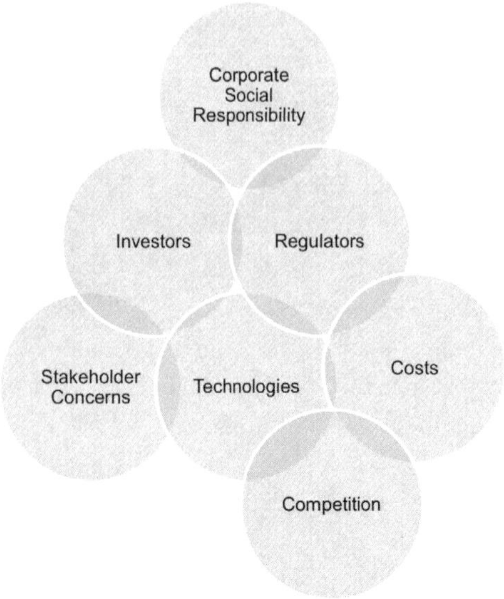

Golder Associates

The Need for Strategic Decision Making Tools

How do I make the right decisions?

How do I develop a water management plan that is sustainable across all business units?

What tools are right for us?

- Decision Tree Tool
- Multi-Attribute Utility Analysis
- Dynamic Simulation
- Sustainability Evaluation

Golder Associates

What is a Decision Tree Tool?

- Developed in Microsoft PowerPoint
- Provides high level guidance through a series of decisions
  - Directs users to detailed corporate documents and regulations
- Covers wide range of production
  - Oil or gas, onshore or offshore, conventional or unconventional
- Can be utilized for all geographic areas
- Includes source options, disposal methods, treatment technologies, vendors

9

Golder Associates

Decision Tree Considerations

Start with high level questions and work down to the details

- What are my sourcing options?
- What are my regulatory requirements?
- What disposal options might I have?
- Do I have to treat?
- What do I have to treat and to what level?
- How do I treat it?
- Who are my approved vendors?

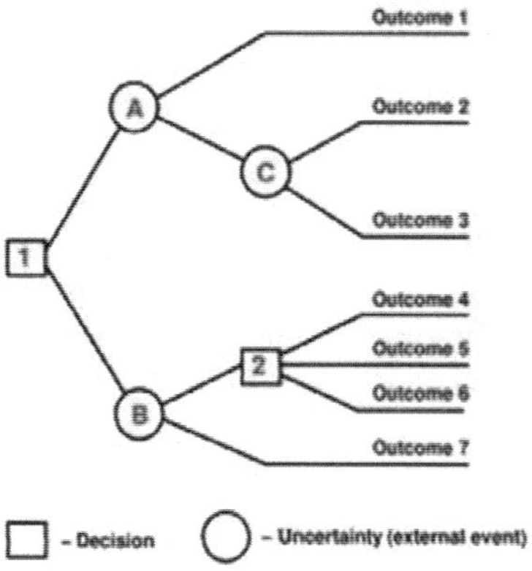

10

Golder Associates

What is a Multi-Attribute Utility Analysis?

- It assigns numerical values
- It involves input from decision makers
- It ranks options based on unique input

Quantifiable Evaluation

Answers the question "What is the Best Choice"?

11

Golder Associates

Multi-Attribute Utility Analysis Steps

12

Golder Associates

Analysis Output

CRITERIA	Produced Water Options					
	Transport to Centralized Wastewater Treatment Facility		Deep Well Injection		On Site Treatment with Reuse	
	Raw Score	Weighted Score	Raw Score	Weighted Score	Raw Score	Weighted Score
1: Physical Attributes	3	12	4	15	3	12
2: Regulatory Requirements	5	17	5	19	4	15
3: Schedule	3	10	4	13	2	8
4: Impacts	4	15	2	9	4	15
5: Cost	2	5	5	13	3	8
Total Score	17.0	59.0	20.0	69.0	16.0	58.0

13

Golder Associates

What is a Dynamic Simulation?

A decision analysis and support tool that…

- Allows you to evaluate, compare and optimize alternatives
- Should be used when the consequences of a proposed action, plan or design cannot be directly and immediately observed
- Should be used when the system under consideration has complex interactions and requires input from multiple disciplines.

Simulation models integrate the various components to better understand the interactions and to enable focusing on the big picture without getting lost in the details

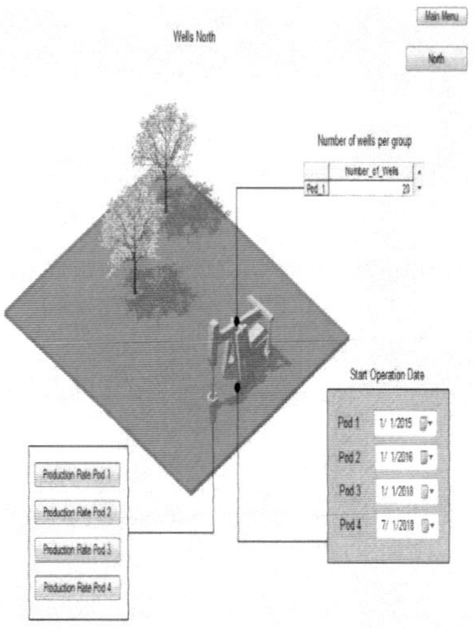

14

Golder Associates

What is a Dynamic Simulation?

A modeling tool that:

- Predicts (forecasts) the future in quantifiable terms (i.e. cost and risk)
- Analyzes different options to understand the long-term implications of management decisions
- Determines consequences of a proposed action, plan or design

It answers the question "What if?"

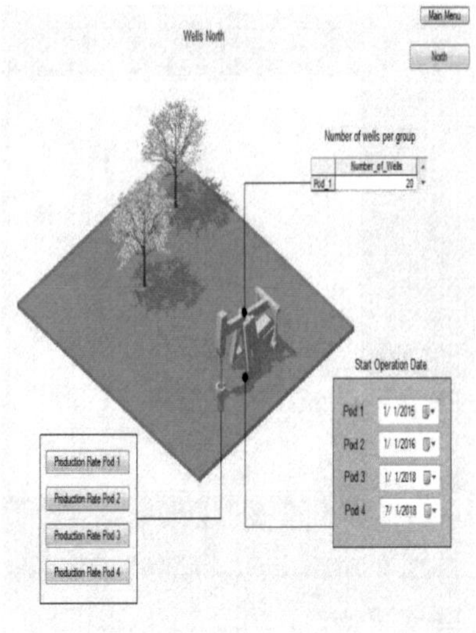

15

Golder Associates

Dynamic Simulation Model

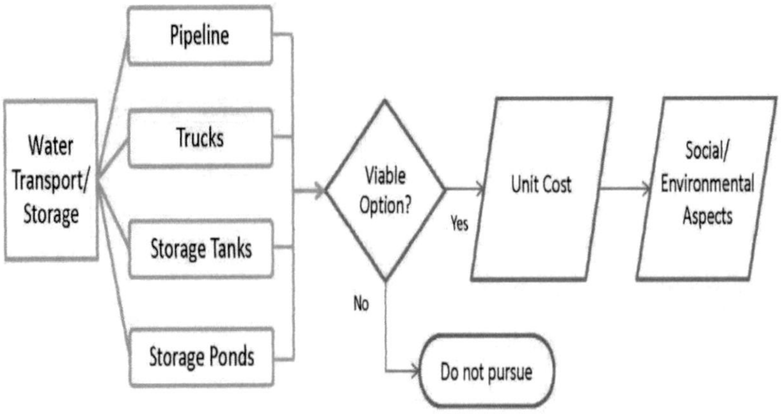

16

Golder Associates

Reasons to use Dynamic Simulation

- Useful for the decision-making process where conflict resolution, negotiation, and dialogue are key
- Helps to understand:
  - how to identify, quantify and manage risk
  - what an optimum water management strategy will be given the level of uncertainty on a project
  - the financial exposure as a result of possible regulatory or operational changes
  - if implementation of a new technology is a good idea

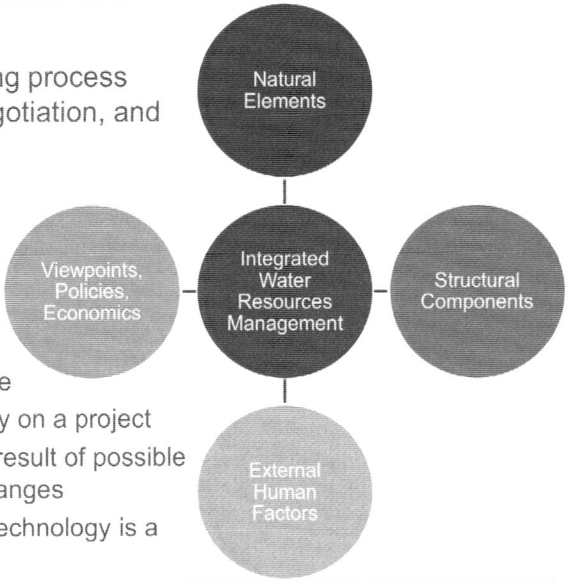

17

Golder Associates

Why Evaluating Sustainability is Critical

UN Global Compact

Global Reporting Initiative™

WBCSD

FIDIC's Guidelines for SD

Equator Principles

Rising Stakeholder Concerns

Multiplication of Principles, Indicators, Guidelines, etc.

Growing Complexity of Sustainability/ Non Technical Risk Issues

How to manage the pressure?
Manage the Business Risk.

18

Golder Associates

Sustainability Evaluation Tool Components

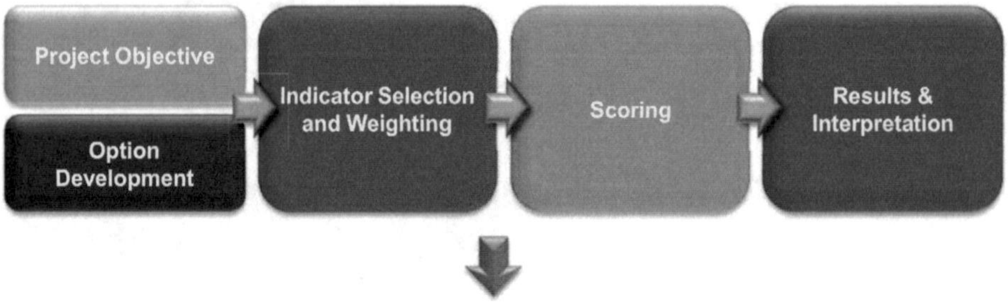

- Looks at the triple bottom line: Environmental, Social and Economic NTRs
- Additional Dimensions -- i.e. Technical, NGO's
- Visual
- Defensible/Transparent
- Quantitative and qualitative
- Organization/Project Specific

Golder Associates

Results & Interpretation

Haul to Central WWTP for recycling/reuse

ENVIRONMENT	66%
SOCIETY	70%
ECONOMICS	67%
TECHNICAL	82%

Onsite Mobile Treatment - on-site frack reuse

ENVIRONMENT	87%
SOCIETY	83%
ECONOMICS	60%
TECHNICAL	82%

Deep Well Injection

ENVIRONMENT	65%
SOCIETY	52%
ECONOMICS	87%
TECHNICAL	75%

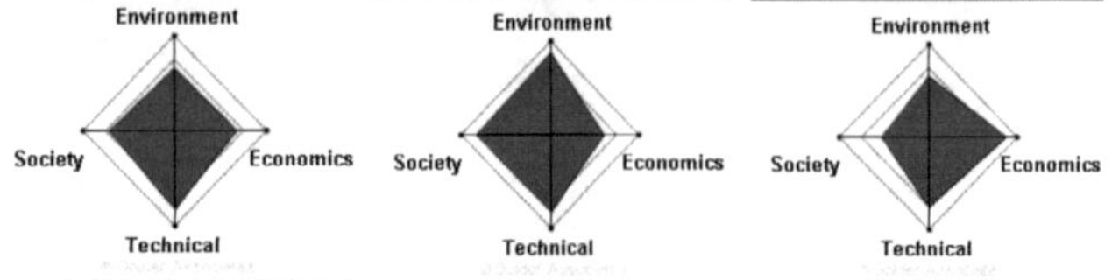

- Identifies option which optimizes environmental, social and economic aspects
- Identifies option which is technically feasible
- Transparent and visual – efficient reporting – better informed decisions

Golder Associates

Conclusions

- Strategic planning is critical for establishing a sustainable, long-term water management plan and reducing project risk
- An effective water management strategy should incorporate all aspects of the water management cycle and all aspects of your business including outside stakeholders
- There are a number of tools that can be used to help develop and formalize a strategy. Tools should be:
  - Relevant for your project
  - Usable with the information and data available
  - Cost-effective in relation to the potential benefit derived from their use
  - Capable of being modified as necessary as the project progresses
  - Suitable to your corporate culture

Thank you!

- Melinda Truskowski – Global Unconventional Gas Leader, Denver
  - mtruskowski@golder.com
  - 720-962-3431
- Tekla Taylor – US Energy Services Leader, Denver
  - ttaylor@golder.com
  - 303-980-0540
- shalegas@golder.com

Golder Associates

AUTHOR INDEX

AIChE
120 Wall Street, FL 23
New York, NY 10005-4020

ISBN 978-1-5108-1597-1

CURRAN ASSOCIATES INC.
proceedings.com